EFFECTIVE INNOVATION

Educated at St Paul's School, John Adair has enjoyed a varied and colourful career. He served in the Arab Legion, worked as a deckhand on an Arctic trawler and had a spell as an orderly in a hospital operating theatre. After Cambridge he became Senior Lecturer in Military History and Leadership Training Adviser at the Royal Military Academy, Sandhurst, before becoming Director of Studies at St George's House in Windsor Castle and then Associate Director of The Industrial Society.

In 1979 John became the world's first university Professor of Leadership Studies at the University of Surrey. He holds the degrees of Master of Arts from Cambridge University, Master of Letters from Oxford University and Doctor of Philosophy from London University, and he also is a Fellow of the Royal Historical Society.

In 2006 the People's Republic of China conferred on John the title of Honorary Professor of Leadership Studies in recognition of his 'outstanding research and contribution in the field of Leadership'. In 2009 the United Nations appointed him to be Chair of Strategic Leadership Studies at its central college in Turin.

www.johnadair.co.uk
www.adairleadershipdevelopment.com

Other titles in John Adair's
EFFECTIVE *series*

EFFECTIVE Communication
EFFECTIVE Decision Making
EFFECTIVE Leadership
EFFECTIVE Motivation
EFFECTIVE Teambuilding
EFFECTIVE Time Management

EFFECTIVE INNOVATION

THE ESSENTIAL GUIDE TO STAYING AHEAD OF THE COMPETITION

JOHN ADAIR

PAN BOOKS

First published 1996 by Pan Books

This edition published 2015 by Pan Books
an imprint of Pan Macmillan, a division of Macmillan Publishers Limited
Pan Macmillan, 20 New Wharf Road, London N1 9RR
Basingstoke and Oxford
Associated companies throughout the world
www.panmacmillan.com

ISBN 978-1-5098-1210-3

Copyright © John Adair 1996, 2009

The right of John Adair to be identified as the
author of this work has been asserted by him in accordance
with the Copyright, Designs and Patents Act 1988.

All rights reserved. No part of this publication may be
reproduced, stored in or introduced into a retrieval system, or
transmitted, in any form, or by any means (electronic, mechanical,
photocopying, recording or otherwise) without the prior written
permission of the publisher. Any person who does any unauthorized
act in relation to this publication may be liable to criminal
prosecution and civil claims for damages.

The Macmillan Group has no responsibility for the information
provided by and author websites whose address you obtain from
this book ('author websites'). The inclusion of author website
addresses in this book does not constitute an endorsement by or
association with us of such sites or the content, products,
advertising or other materials presented on such sites.

A CIP catalogue record for this book is available from
the British Library.

Typeset by Setsytems Ltd, Saffron Walden, Essex

This book is sold subject to the condition that it shall not,
by way of trade or otherwise, be lent, re-sold, hired out,
or otherwise circulated without the publisher's prior consent
in any form of binding or cover other than that in which
it is published and without a similar condition including this
condition being imposed on the subsequent purchaser.

Visit **www.panmacmillan.com** to read more about all our books
and to buy them. You will also find features, author interviews and
news of any author events, and you can sign up for e-newsletters
so that you're always first to hear about our new releases.

'He that will not apply new remedies must accept new evils: for time is the greatest innovator.'

Francis Bacon

CONTENTS

Foreword ix
Introduction xi

PART ONE: THINKING TO SOME PURPOSE
1 Creativity and innovation 3
2 How the mind works 19

PART TWO: SEVEN HABITS OF SUCCESSFUL CREATIVE THINKERS
3 Habit One: Going beyond the nine dots 39
4 Habit Two: Welcoming chance intrusions 61
5 Habit Three: Listening to your depth mind 68
6 Habit Four: Suspending judgement 80
7 Habit Five: Using the stepping stones of analogy 87
8 Habit Six: Tolerating ambiguity 98
9 Habit Seven: Ideas banking 111

PART THREE: MANAGING FOR INNOVATION
10 How to manage innovation 133
11 The innovative organization 159
12 The art of brainstorming 177
13 Taking good ideas to market 191

Appendix: Solutions to problems 206
Acknowledgements 211
Index 213

FOREWORD

Welcome to this fully revised and updated edition of *Effective Innovation*. The underlying practical philosophy of this book, as well as its principles, models and frameworks, have withstood the test of time. I am delighted, therefore, to be able to share them with you here.

I suppose that as individuals we should all be – or be ready to be – innovators in our personal lives. My focus in this book, however, is on those of you who are called upon to be leaders in organizations, leaders at all levels – team, operational or strategic.

In one of my earlier books I wrote that 'change throws up the need for leaders and leaders bring about change'. Leadership is about taking people with you on a journey, which is another word for change, and that almost invariably involves new ideas and new ways of doing things. Are you ready for that?

The constant demand for innovation can be daunting, and I can understand if you feel it to be so. But, remember, you personally don't have to have all the ideas or do all the work. Creativity will be widely distributed throughout your team or organization, although you will always find that some individuals are more creative than others – that is life.

It takes effective teamwork to bring creative ideas to market in the form of improved, or new, products and services. Building that kind of teamwork, with its wide range of know-how, talents and skills, calls for effective leadership on your part.

The good news is that this book will show you how to both nurture creativity and harness it with teamwork so that it produces all the innovation you need. As the case studies show, it has been done before. Now it's your turn. Good luck.

John Adair, 2009

INTRODUCTION

'What a wonderful thing the human brain is,' said one manager a few decades ago. 'It starts working when you wake up and stops as soon as you get to the office.' Nothing could be further from the truth today.

Ideas about work have undergone a revolution in recent times. Gone are the days when an employer could hire people's physical energies without their minds. Today, success at work – and longer-term employability – depends largely upon mental contribution. Most of us are 'knowledge workers' now. If our brains aren't fully engaged by the time we reach work – and throughout the day – our career prospects are limited!

Creative thinking, or having new ideas, is becoming an ever more important part of work. But innovation is more than personal creativity. It is *the process of taking new ideas through to satisfied customers*. And it occurs at every stage in any business. Potentially it involves the whole team at work. Which is why innovation ought to be an essential part of your business strategy.

Effective innovation has three overlapping dimensions. You can picture them like an oyster shell:

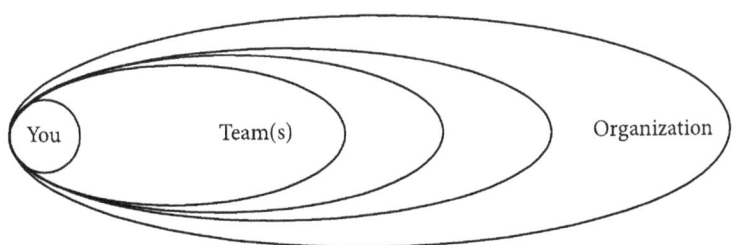

By you, of course, I mean you as an individual. And in this book *you* and *your skills* will always come first. Why? Because new ideas are the pieces of grit in the organizational oyster than can grow into white lustre pearls. Organizations don't have new ideas. Teams don't have new ideas – only *individuals* have new ideas. That's why you come first.

But taking ideas to market involves teamwork, for one person cannot do it on their own. *Innovation requires teams*. Notice that I have put several teams on the shell, because you may well belong to several: core department or functional team, project group, quality circle, executive management team or board of directors.

The *organization* you work for encompasses all its parts – including you and the teams you belong to – but it's more than the sum of its parts. It has a life of its own. In *Effective Leadership* (1989) I suggested that all organizations develop a *group personality* as well as possessing *three overlapping areas of need*:

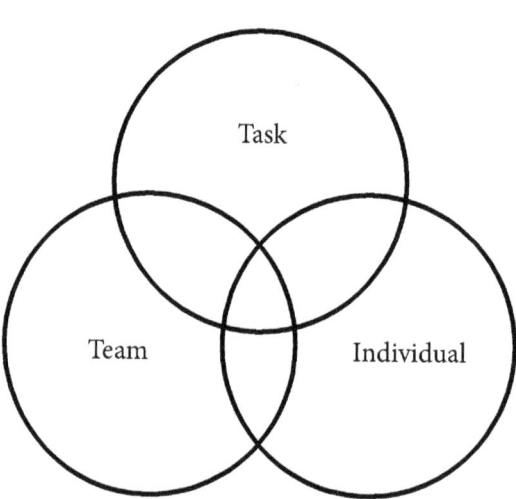

That *group personality*, or culture, as it is commonly called, plays a vital role in innovation. Some cultures encourage innovation; others stifle it. It's not much good being a creative individual and building creative problem-solving teams if your parent organization poses a hostile environment to new ideas. This book encompasses all three dimensions, or layers, of the outer oyster shell.

The aim of this book is quite simply to help you become a more effective innovator. Obviously I cannot do it for you – you must learn how to do it for yourself. But that is no bad thing. As Winston Churchil once said to a friend, 'I don't mind learning, but I hate being taught.'

I am assuming that you have a direct interest in being able to produce new ideas and bring them to market. You may work for an organization which is in the throes of change. Or you may be self-employed in a business that demands fresh thinking and an innovatory approach. Either way, creative thinking and innovation are important to you.

Within the overall aim of the book there are three specific areas of need that you will need to focus on to become an effective innovator. With this in mind I have divided the book into three parts, each with its own area of need.

What do you need to become an effective innovator?

- **You need a conceptual framework for *understanding* how your mind works when it seeks new ideas.** Part One clarifies the concept, provides some background, and offers you a model of your mind at work.
- **You need to develop your *skills* as a creative or innovative thinker.** Part Two describes seven habits of successful creative thinkers, showing how they can become *your* habits of mind. Each of them has a chapter to itself.
- **You need to be able to *manage* innovation.** Part Three encompasses what you have to do to help teams develop new ideas, the ways in which you can build an innovative organization, and how to take your creative ideas into the market place.

HOW TO USE THIS BOOK

This is not the kind of book where you have to start at the beginning and read the chapters in sequence through to the end. You may have a natural preference to jump forward and backwards. Feel free to do so. The chapters are designed to be more or less self-contained, so you can read them when you have time. Imagine the chapters as being, not steps in a ladder but, rather, spokes in a wheel. You don't have to be a step-by-step reader to enjoy this book!

As with the other books in this series, it is important to understand the process of learning. I suggest that you will learn little or nothing about creativity and innovation unless you make a conscious effort to relate the general principles to your real-life experience. It is essential to bear in mind that we learn by the interaction of:

PRINCIPLES   EXPERIENCE
   or   and   or
 THEORY   PRACTICE

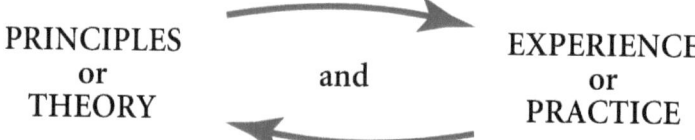

It is when sparks jump between these two poles – the general and the actual – that learning occurs. So you need both. The various case studies and examples in this book are designed to be stepping stones:

PRINCIPLES → THIRD-PERSON EXAMPLES → YOUR EXPERIENCE

Equally the process must work in reverse. Your practical knowledge, gleaned from both the observation of actual creative thinkers or innovators and your own practical experience, must be brought to bear in a constructively critical way on the ideas presented here.

INTRODUCTION xv

Checklists have been included to help you relate the ideas of this book to your own particular needs, problems and opportunities. The more time and thought you invest in answering the questions in them, the more useful this book will be.

USING THE BOXES

As a small anthology on creative thinking and innovation, I have introduced some material in boxes to supplement the main body of text. My main aim in doing so is to stimulate further thought. Reading the boxes may be a further exercise you could set yourself should you choose to return to the book a second or third time.

Finally, this isn't a textbook, but more of a practical guidebook. I want you to find it stimulating and enjoyable, as well as being instructive, so I've added some puzzles and exercises to make the whole experience more fun.

If you do choose to invest some time and mental effort into reading and working on the following pages, they should help you to:

- Develop your understanding of the creative process
- Overcome barriers or blocks to having new ideas
- Enlarge your parameters of vision
- Learn to build on ideas as well as criticize them
- Increase your tolerance for uncertainty and doubt
- Listen, look and read with a creative attitude
- Make time to think
- Become more confident in yourself as a creative person
- Be a more effective leader or member of innovative teams
- Know how to manage innovation in your organization

PART ONE

THINKING TO SOME PURPOSE

The concepts of creativity and innovation overlap considerably, but they are not the same. What they have in common is the idea of *newness*. A creative idea is a new idea, at least to the person who has it. It may involve an act of synthesis uniting two or more apparently disparate ideas or things.

There isn't a special box in your brain labelled 'Creative Thinking Department'. When it comes to creative thinking and innovation the whole brain is involved. One part or another may be at work at certain phases, but any thinking – even creative thinking – always uses the brain as a whole.

By the end of reading and studying Part One you should have:

- A clear **understanding** of the meaning of the concepts involved, together with some of the better known contemporary theories of creativity, such as lateral thinking.
- A simple but comprehensive **sketch map** of how your mind works when it's thinking to some purpose.

- Some understanding of the **fields or activities** in which you are most likely to become an effective creative and innovative thinker.
- An introduction to a central paradox of creative thinking, namely that it's both a supremely **solitary** craft and yet also intensely **social** in nature. You can't think for long on your own.

1
CREATIVITY AND INNOVATION

A distinguished visitor to Henry Ford's auto plants once met him after an exhaustive tour of the factory. The visitor was lost in wonder and admiration. 'It seems almost impossible, Mr Ford,' he told the industrialist, 'that a man, starting twenty-five years ago with practically nothing, could accomplish all this.'

'You say that I started with practically nothing,' Ford replied, 'but that's hardly correct. Every man starts with all there is. Everything is here – the essence and substance of all there is.'

The potential materials – the elements, constituents or substances of which something can be made or composed – are all here in our universe. We humans cannot make anything out of nothing. If you narrowly define creativity as 'to bring a thing into existence without any previous material at all to work on', as Thomas Aquinas did, then clearly it is 'only God who creates – Man rearranges'.

WHAT IS TRUE CREATIVITY?

Clearly we cannot create in that divine sense of making something out of nothing. Some people, however, like to confine the concept of creativity to the 'creative arts' which most approach that supernatural ability. They reserve the word 'creative' for products that are very far removed from the original raw materials used.

A masterpiece by Rubens was once a collection of blue, red, yellow and green 'worms' of paint on the artist's palette. Now the physical materials in this type of creative work – paints and canvas for an artist, paper and pen for an author – are entirely secondary. Creation here is more in the mind. Perception, ideas and feelings are combined in a concept or vision. Of course, the artist, writer or composer needs skill and technique to translate what is conceived in their mind on to canvas or paper. Such artists are often thought to be inspired by God, or some spirit or genius outside themselves, that enables them to transcend normal bounds in creating works of truth or beauty.

At the other end of the scale, the concept of creativity is sometimes stretched so far as to almost empty it of any meaning. You may think of an idea as being creative, for example, because it is new to *you*. But if this is the only criterion used, it is on the bottom rung of the ladder.

> Spencer Penknives, a long-established Sheffield cutlery company, was going through hard times. Henry Parker, the managing director, came into the office one morning and summoned his senior executives. 'I was doing some creative thinking in the bath last night,' he said, 'and I've had a good idea.' An air of expectancy suddenly developed. 'We must ask our main customers why they aren't buying our products in such quantity now.' A stunned silence followed. After the meeting on the way out, the sales director commented to a colleague, 'Some creative idea! We've been saying that for the last two years.'

Having an idea that is new to you is certainly a step in the right direction. But you must also ask yourself the following: Is it new to others? Has it been brought to market already?

Now in truth it is very rarely possible to have a truly new idea. *Homo sapiens* have been on earth for a very long time. So almost anything you can conceive has probably been thought of or invented by someone else before.

Did you know that . . . ?

- Solomon's Temple was protected by lightning rods
- Emperor Nero devised a slot machine
- Another Roman emperor had three elevators in his palace
- Hindus used the cow-pox virus centuries before Jenner
- The reaping machine was described as a 'worn-out French invention' in the sixteenth century
- A thousand years before Christ, the Chinese extracted digitalis from living toads to treat heart disease, recorded earthquakes undetected by the human senses, and developed an instrument that always pointed north – the compass

You can see what the author of *Ecclesiastes* meant when he coined the phrase 'there is nothing new under the sun'.

But some ideas may be born before their time. The Caesars, for example, may have had a few elevators, but the state of technology of their times meant that they could not be produced on any scale. Leonardo da Vinci may have sketched helicopters and submarines but, again, the technology for making and powering them wasn't available.

Originality, *per se*, then, is not enough. What is more important is that an idea should be relatively new and feasible *in your given situation or environment*. For example, it's too late for you now – just as it was too early for Leonardo – to make your personal fortune from inventing the helicopter. That idea and its technology are no longer new.

Notice, incidentally, that whenever the tide of technology advances – especially if it makes a quantum leap like the advent of the internet – a whole range of 'old'/new ideas in the background of our collective memory as the human race suddenly becomes feasible.

DEFINITIONS

Why do people *perceive* anything as new? They obviously won't do so if it's already very familiar. That may give us some clues about innovation. The following questions point to the perception of newness in an idea or concept, product or service:

- Has it recently come into existence?
- Has it been made or used for only a short time?
- Is it freshly made and unused?
- Is it different from something similar that has existed before?
- Is it of superior quality?
- Has it just been invented, created or developed?

You can now see that new is not the same as original, which, strictly speaking, applies only to something that is the first of its kind to exist. As our knowledge is limited it's difficult to establish a claim to true originality. So often research or enquiry will unearth someone who has thought of it before. Nor does new always have the extra overtones of novel, which adds to newness the nuances of the strange and the unprecedented.

You may find it useful to think of successful innovation as being comparable to the production of a play or film in which different people – say, a director, producer, author and scriptwriter – are involved. In successful innovation one or more of the following individuals will be at work:

KEY PLAYERS IN INNOVATION	
ROLE	NOTES
Creative thinker	Has the power or quality to produce new ideas, especially ones not known to have existed before.
Innovator	Can bring in or introduce something new or as new, such as a product or service to the market. Also, alters or makes changes to an established product or service.
Inventor	Comes up with a new and potentially commercial ideal. Often combines both creative thinker and innovator.
Entrepreneur	Conceives or receives ideas and turns them into business realities. Often uses OPB (Other People's Brains) and OPM (Other People's Money) to develop a market opportunity.
Intrapreneur	'Takes hands-on responsibility for creating innovation in any kind of organization. The intrapreneur may be the creator or inventor but is always the dreamer who figures out how to turn an idea into profitable reality.' Gifford Pinchot, *Intrapreneuring: Why You Don't Have to Leave the Corporation to Become an Entrepreneur* (Harper and Row, 1985)
Champion	Picks up an idea, not necessarily his or her own, and runs with it. Shows commitment and tenacity in seeing it developed properly and successfully implemented.
Sponsor	Gives an idea the backing it deserves. Usually a senior manager who believes in it and influences key people to clear the way and help overcome obstacles as it is taken to realization.

CASE STUDY: GOOGLE – THE WORLD'S MOST POWERFUL INTERNET SEARCH ENGINE

What made Google the fastest growing company in the history of the world? The story began with two creative-thinking self-confessed computer nerds at Stanford University; Larry Page and Sergey Brin. Together they invented a convincing answer to the question that is in the mind of every internet user: 'How can I find, in order of importance, the web pages that are relevant to my present concern?'

A core ingredient of their solution was a program they developed called PageRank, which signals the importance of any given web page by counting the number of other pages linked to it. But the formula they used to 'score' the relevance of a given website blended in other criteria as well: how often key words appear on a given page; whether the website home appears in the page's title; and so on. Armed with these insights, the innovative partners founded the company in a garage in 1998 and based the name on a misspelling of the word for the number 10^{100}: the googol. Its stated mission was 'to organize the world's information and make it universally accessible'.

Equally innovative was Google's business strategy. No money was spent on advertising: since the site was both incredibly useful and free, its promotion took place entirely by word of mouth. Meanwhile, Page and Brin were raking in money through a supremely simple device: the 'sponsored links' section on the right of each search page. Every time you click on any of the 'AdWords' in that space, you're taken to the website of the company which has bid for the privilege of 'owning' those precise 'AdWords'; and the company pays Google for each click. The small ads are hugely profitable: over a six-month period they earned Google some $2.6 billion.

The future of Google – as with all companies – depends on creativity and innovation, and it has extended these principles to the way the company is run. Google's philosophy is to give its highly-paid staff free rein. Employees are urged to devote 20 per cent of their time to their own pet projects. Hence the development of Google News (a global news engine which searches national and regional newspapers); Froogle (now Google Product search – a price comparison site); Google Talk (a way of making free phone calls over the net); and Google Book Search (a scheme to make the full text of the world's books searchable by anyone).

CASE STUDY: EDWARD DE BONO

A good example of an effective innovator is Edward de Bono, who introduced the concept of lateral thinking in his book *The Use of Lateral Thinking* (Jonathan Cape, 1967). What in fact he did so successfully was to reintroduce the old concept of creative thinking as new into the market, by giving it a new name and a new gloss of paint.

Born in Malta in 1933, Edward de Bono went to Oxford University as a Rhodes Scholar. After obtaining degrees in medicine, psychology and physiology he became a university teacher and subsequent author of some forty books. He describes himself as 'a thinker about thinking'. So what does he have to teach us?

De Bono's best known book remains his first one. In it he introduced in a vivid way his key distinction between logical or vertical thinking on the one hand and lateral thinking on the other:

> Logic is the tool that is used to dig holes deeper and bigger, to make them altogether better holes. But if the

hole is in the wrong place, then no amount of improvement is going to put it in the right place. No matter how obvious this may seem to every digger, it is still easier to go on digging in the same place than to start all over again in a new place. Vertical thinking is digging the same hole deeper, lateral thinking is trying again elsewhere.

In logical or vertical thinking one 'arrangement of information develops directly into another'. It is a step-by-step procedure. For example, when solving a mathematical problem or in arguing from a logical premise you are using vertical thinking. Much of our day-to-day reasoning does move – or attempt to move – from one set of information to another in a chain.

Lateral thinking, by contrast, involves abandoning the step-by-step approach altogether. 'Lateral' comes from the Latin word *laterus*, meaning 'a side'. So lateral thinking is thinking to one or other side. The phrase when Edward de Bono introduced it was undoubtedly new to his audience. As is so often the case, however, the concept was probably not new. (The French philosopher and thinker, Etienne Souriau, had said some time before that *'pour inventer il faut penser à côté'* – 'to invent you must think aside'.)

> One of my early cars developed a leak and the hollow bodywork tended to fill with water. Several expensive visits to the garage failed to solve the problem. A manager I met on a course looked at it for me. 'Why not bore a small hole in the bottom and let the water out?' he suggested. It was a good example of lateral thinking. I had seen the problem in terms of how to *keep out* the water. Suddenly I saw that the solution – or at least part of it – lay in how to *let out* the water. Although it was not a complete solution it did solve my immediate problem.

As this story suggests, what lateral thinking does is to encourage you to rearrange or restructure a problem, in this instance by reversing it. Thus, 'How can I keep the water out?' becomes How can I let the water out? 'As the modern proverb says,' 'A problem is a solution in disguise.'

Henry Ford's innovation of the assembly line provides another example of lateral thinking. The traditional method of car manufacture involved men working down a line of stationary cars. But Ford thought: Instead of men moving to and from the cars, why not move the cars in front of the men? It was a simple but revolutionary idea.

So instead of pursuing a course of action or solution path that isn't working, look to the side and see if there isn't an alternative approach waiting in the wings – the lateral thinking approach. Try the logical approach first and if that doesn't yield results, put on your lateral thinking hat.

The phrase 'lateral thinking' became so widely used that it entered the *Oxford English Dictionary*, defined as 'seeking to solve problems by unorthodox or apparently unorthodox methods'.

Lateral thinking in this original sense is therefore a technique for tackling problems that do not seem to be yielding to the frontal assault of logical or rational thinking. It's particularly useful where all the information is available but the problem needs restructuring or reframing in order to allow the solution to come forwards. Following a military analogy, it is trying to outflank the problem by moving to one side or behind it, so that it surrenders without a fight.

Look at any object that is clearly in focus and you will notice that around it there is a background that is more or less out of focus. What Edward de Bono in effect suggested is that, if you are stuck, you should transfer your attention from the field of your focus to the ground around it, which might yield a solution. It involves reversing the perception of field and ground.

The case of a helpful statue

The essence of lateral thinking is shifting your attention from the foreground to the background in search of a solution to the problem.

Soichiro Honda was an engineer who excelled in creative thinking and innovation. While he was building his first four-cylinder motorcycle he gradually realized that although the engine was fine, his designers had made the machine look squat and ugly. While pondering the problem, he decided to take a week's break in Kyoto. One day, sitting in an ancient temple, he found himself fascinated by the face of a statue of the Buddha. He felt that he could see a resemblance between the look of the Buddha's face and how he imagined the front of the motorbike would be.

Having spent the rest of the week studying other statues of the Buddha in Kyoto, Honda returned to his factory and worked with the designers to produce a harmonious style that reflected something of the beauty he had glimpsed in Kyoto.

De Bono introduced his lateral thinking as an updated or rephased version of creative thinking, which he called 'a more self-satisfied name' for what he was talking about. Now lateral thinking, as defined above, and creative thinking are by no means the same thing. Lateral thinking is essentially a problem-solving technique or useful habit of mind, whereas creative thinking is a much wider concept. But there is some overlap between the two concepts, for lateral thinking can help you to escape from fixed ideas and generate new ones.

Why was lateral thinking such an effective innovation?

Apart from the fortunate timing of its 'launch', it had two characteristics that 'sold' it in the market place – (remember, innovators have to sell ideas as well as have them):

- It worked on the basis of reversal, or contrast – something that the human eye and brain perceive very clearly. If you can make something black or white, either/or, it is more likely to grab attention. Lateral thinking is also an excellent teaching device. The distinction between vertical and lateral thinking falls into the same category. The necessary over-simplification can be corrected and retrieved later.
- It had none of the fearsome overtones that creative thinking has acquired (see page 4 above). It was not associated with an elite of famous inventors. It looked like a technique or skill anyone could acquire, like playing golf. You don't have to be born with a lateral thinking mind – it's more of a game. 'As in golf,' wrote de Bono, 'a sort of general training would be of some use.'

In his subsequent spate of books, Edward de Bono capitalized on the brand name of lateral thinking and its suggestion of instant creativity. He widened the product to include creative thinking proper, so that lateral thinking came to mean 'the generation of new ideas and the escape from old ones'.

A lateral thinking approach that restructures a question or problem can pay great dividends in many different fields. In the quest to understand leadership, for example, perhaps the single most important step has been to turn away from researching the leader (and his or her personality traits or qualities) to studying the group. The concept of the three overlapping areas of need present in working groups or organizations – task, team and individual – led naturally to the idea that leadership should meet these areas through providing necessary functions. This includes defining objectives, planning, setting and maintaining group standards and encouraging, motivating and developing each individual. This switch of emphasis from leadership *qualities* to leadership *functions* has proved to be immensely fruitful in the fields of selection and training.

A similar piece of lateral thinking has also taken place in the field of creativity and creative thinking, turning attention away from what makes individuals creative (not unlike the study of leadership qualities) to a different question: What *prevents* people from being creative or innovative? Obtaining lists of blocks, barriers and obstacles – in individuals, groups and organizations – stems from reversing the problem in this lateral way.

Notice the hidden assumption that is now smuggled in; namely that people are naturally creative. As I said in the introduction, I have no means of knowing whether or not this is true. But I am completely persuaded that the vast majority of people are at least potentially capable of contributing effectively to the innovative process. There is a great deal of circumstantial evidence from industry and commerce, not least in the aspect of total quality improvement, to support that view.

Consequently, a strategy of seeking to *reduce* the negative or inhibiting factors which work against creativity or innovation may be the best one to adopt. Use yourself as a live experiment. Can you identify any of the following in yourself?

SOME BARRIERS TO CREATIVE THINKING	
Negative attitude	A tendency to focus on the negative aspects of problems and expend energy on worry, as opposed to seeking the inherent opportunities in a situation.
Fear of failure	A fear of looking foolish or being laughed at. Yet Tom Watson, founder of IBM, said: 'The way to accelerate your success is to double your failure rate.' Failure is a necessary condition of success.
Executive stress	Not having time to think creatively. The over-stressed person finds it difficult to think objectively at all. Unwanted stress reduces the quality of all mental processes.
Following rules	Some rules are necessary but others encourage mental laziness. A tendency to conform to accepted patterns of belief or thought – the rules and limitations of the status quo – can hamper creative breakthrough.
Making assumptions	A failure to identify and examine the assumptions you are making to ensure they are not excluding new ideas. Many *unconscious* assumptions, in particular, restrict thinking.
Over-reliance on logic	Investing all your intellectual capital into logical or analytical thinking – the step-by-step approach – can exclude imagination, intuition, feeling or humour.

The biggest barrier to creative thought, however, is believing you are not creative. A legacy often from poor teaching at school, this barrier really stops people in their tracks from even entering the race. It is in fact a restriction or assumption that we impose on ourselves. Whatever your self-image may or may not have been in the past, it is an unproven hypothesis as far as the future is concerned.

Remember that creative thinking is natural to *Homo sapiens* as a whole. Just look around and you will see evidence of human creativity and innovation everywhere. Why not assume that some of that creativity is present in you? Instead of trying to develop your creative ability, concentrate on removing the barriers, dams or blocks that prevent your mental energy from producing new ideas, new ways of working at things.

This lateral thinking approach is good at getting us to identify the forces or factors that work *against* creative thinking and innovation. But soon, like a patient beginning to feel better, we want more fuel – the positive principles or skills that can guide us on the next stage of the journey described and discussed in Part Two.

But first we must look at how the mind works in the next chapter. Any form of thinking is best done by working *with* the grain of your mind rather than against it, and you can only do this by *understanding* your own mind. Doing this will, of course, then help you to understand how other people's minds work – for we are all an analogy of others. You can then work in concert or harmony with a team of minds on a problem or opportunity.

KEY POINTS: CREATIVITY AND INNOVATION

- Men and women don't create from nothing. We tend to reserve the word 'creative', however, for products that are far removed from the original materials used. It usually implies a value judgement.
- The concept of *newness* is a relative one: what may be new in one context is old in another.
- An innovator is someone who introduces something new

or as new, such as a product or service to a market. The concept includes improving an *existing* product or service.
- Lateral thinking, itself a good example of an innovation, is a good lead into the differences of emphasis between logical and 'vertical' thinking and the more creative exercise of the mind.
- Using that approach, it's not a bad idea to switch your attention away from how to *be* creative to what *prevents* you from being creative, such as negative attitudes, fear of failure, executive stress, following rules, making assumptions and over-reliance on logic.
- Lastly, rid yourself of one pernicious block to creativity: the belief that you are not a creative person. In fact, *creativity* is much more widely distributed than was previously imagined. Creativity is something everyone can learn and this book will show you how to develop it.

CHECKLIST:
CREATIVITY AND INNOVATION

	Yes	No
Answer Yes or No to the following:		
Can you give at least one example of when you have adopted a lateral thinking approach?	☐	☐
Do you tend to think positively about problems and see the opportunities they often disclose?	☐	☐
Have you ever held back an idea or suggestion for fear of being ridiculed?	☐	☐
Do you tend to criticize yourself severely?	☐	☐
Are you too busy to think?	☐	☐
Would you say that your stress level is above average in your line of work?	☐	☐
Would your partner or best friend describe you as a conformist – an other-centred person who puts acceptance by the group above all other values?	☐	☐
Have you ever felt so inspired by a new idea that you had to telephone a friend just in order to share it?	☐	☐
'All problems in life can be solved by logic.' Would you agree with this statement?	☐	☐
'Many more people are capable of creative thinking. There is still a vast unused intellectual potential in most organizations today.' In your opinion, is this an accurate statement?	☐	☐

HOW THE MIND WORKS

The brain and the mind are not the same thing. Imagine a television set: what appears on the screen is the mind at work; what you see when you open up the back and look inside – that complicated mass of wiring – is the brain. Obviously, as we all know, the mind is related to the brain. Some aspects of mental functioning are located in specific parts of the physical brain. But the overall relationship between mind and brain – especially the phenomenon of consciousness – contains many mysteries. Neuro-scientists study the brain; philosophers and psychologists study the mind.

SPLIT BRAIN THEORY

In 1981 Roger Sperry received the Nobel Prize for developing the split-brain theory. According to Sperry, the brain's two hemispheres have different but overlapping functions. The right and left hemispheres of the brain both specialize in particular kinds of thinking processes.

In general, in 95 per cent of all right-handed people, the left side of the brain cross-controls the right side of the body and is responsible for analytical, linear, verbal and rational thought. In most left-handed people the hemispheric functions are reversed.

Left hemisphere

Logic
Sequential
Verbal
Linear
Analytical
Reasoning
Explicit
Calculation

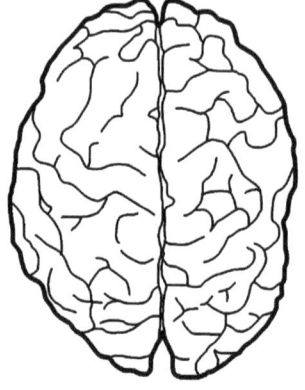

Right hemisphere

Intuition
Images
Visual
Spatial
Creative
Holistic
Colour
Emotion

Split-Brain Theory

It is a left-brain function you rely on when doing your accounts, recalling names and dates or solving logical problems.

The brain's right hemisphere controls the left side of the body and is the source of imaginative, non-verbal and artistic thinking. Whenever you think of a place, become engrossed in a painting or sculpture, imagine, or simply daydream, you are engaging in right-brain function. At school, right-brain processes are often held in less regard than logical, analytical thinking.

Within the practical context of this book, what we are really concerned with is how the *mind* works, not the brain as such. Brain research is certainly interesting to many people but as yet there is no evidence that the knowledge it yields is of any use in increasing our intelligence. What it *does* do, however, is to remind us of how under-used the resources are that we have in our brains. After all, we each have an estimated ten billion brain cells – more in number than the current world poplulation, or all the leaves in the rainforests of South America. Each brain cell can connect with about 10,000 of its neighbours, giving 1 plus 800 noughts of combinations.

Of course, no model of the mind can ever be complete. My mind or yours can create a model or picture of its own workings just as we can build a model of our DNA with coloured bricks. But it can't put into it the universe of all mental activities or states. We can only select with a specific purpose in view. General theory eludes us.

A THEORY OF EFFECTIVE THINKING

My own model of the mind, introduced in *Training for Decisions* (1969) and developed further in *Effective Decision Making* (1985), suggests that it has three meta-functions that interlock like a jigsaw puzzle, as seen below.

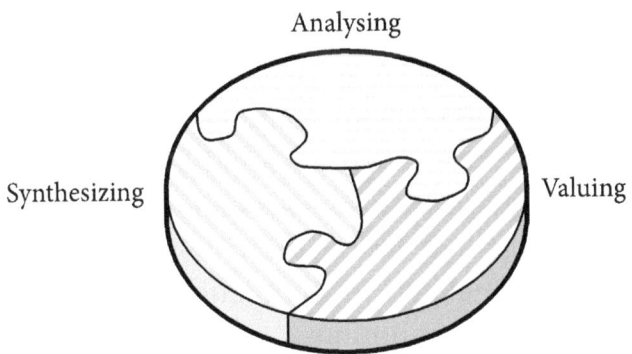

Meta-Functions of the Mind

The main suggestion of this model is that thinking, in the deliberate or purposeful sense of the word, takes three basic forms, all stemming from the evolutionary story of humankind in response to its environment. These basic forms are: analysing, synthesizing, and valuing.

- *Analysing* comes from the Greek verb meaning 'to loosen'. Its primary meaning is 'resolving into simple parts'. In other words, when I take my watch to pieces I can strictly be said to be analysing it. The word, however, has overtones of meaning beyond this simple physical act. Indeed, the concept of 'loosening' does not imply a complete separation of elements. The knot tying two pieces of rope may be untied, or merely loosened, so that the nature of the knot can be understood. Analysis implies the tracing of things to their sources, and the discovery of general principles underlying concrete phenomena.
- *Synthesizing* again comes from Greek. The opposite of analysing, it is 'the putting together of parts or elements so as to make up a complex whole'. Indeed the Latin verb *cogito*, 'I think', can be derived from roots meaning 'to shake together'. When the resulting whole is substantially new – especially if it is original – we can describe the synthetic process as creative.
- *Valuing* is the third meta-function of thinking. It is not finally reducible by or to analysis, or to synthesis, or any combination of them. Valuing, or thinking in relation to values or standards, should take its place beside analysing and synthesizing as a major form of thinking in its own right. It lies at the core of judgement and places an essential note in all situations involving choices or decision between options.

These meta-functions work together in our thinking. We are not usually conscious of the gear changes. Yet the balance between them is changing from moment to moment. So one minute you may be primarily analysing and the next valuing. They are complementary. This point can often be appreciated more clearly when we see the distortion that occurs when one mode of thinking becomes dominant at the expense of others, either in a particular individual or group of people. A group of academics, for example, may become over-analytical or too ready to give negative criticism. In all effective thinking – including the drama of creative thinking – all three meta-functions are involved in a dynamic trio, although one actor may be on stage while the other two are waiting in the wings.

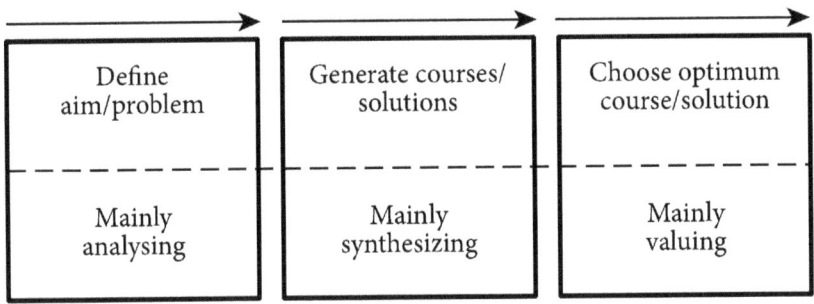

To take a *group* across the river you must build a bridge with three pillars. As you'll see, these pillars draw mainly on one of the meta-functions, so it helps to be able to separate them.

Crossing the River

THE DEPTH MIND PRINCIPLE

The title of this section is taken from the metaphor of the sea. Instead of an apparent dichotomy between two compartments of the mind (conscious and unconscious) this image suggests a continuum between 'surface' and 'depth' minds. Imagining the sea allows us to see the light of consciousness penetrating much further from the surface into the 'caverns of the mind' and gradually becoming dimmer in the depths of 'no-man-fathomed'. The depth mind may be involved in all of the three modes of thinking – analysing, synthesizing and valuing. Thus we have two inter-related variables:

- The alternations and interactions of analysing, synthesizing and valuing
- The constant automatic changes of gear between different levels of the mind, from surface to depth, through which each of the meta-functions moves

The three meta-functions, then, work on different levels of consciousness in the brain, possibly at the same time. Synthesizing, for example, can be done on a conscious level, as when assembling an electric plug or making a toy. But the process of putting together parts or elements so as to make a complex whole can also take place at pre-conscious, semi-conscious or unconscious levels.

> *Dust as we are, the immortal spirit grows*
> *Like harmony in music; there is a dark*
> *Inscrutable workmanship that reconciles*
> *Discordant elements, makes them cling together*
> *In one society.*
>
> William Wordsworth

Many creative people like the poet Wordsworth bear witness to the fact that the welding of apparently disparate or diverse 'parts' into a pattern which is new (in the sense that the resultant whole has not been known before) takes place in the unconscious mind. The flash of inspiration, the sudden bright idea, the Eureka experience, is often the result of a period of subliminal mental activity.

Bisociation

Arthur Koestler in *The Act of Creation* (1964) coined the new word 'bisociation' to describe the essence of creativity as *putting together two unconnected facts or ideas to form a single*

idea. The experience of bisociation usually releases tension. There is a spark of illumination, producing a cry of 'Eureka!' or at least 'Aha!'. As Koestler remarks, it is like the release of laughter after the unexpected end of a joke – the 'haha' reaction. Christmas cracker jokes sometimes turn on bisociation: 'What do you get if you pour boiling water down a rabbit hole? A hot cross bunny.'

Since people first became more aware of the unconscious mind largely through the influence of Freud, it has become rather typecast. The *conscious* is often seen as the seat of reason and order. The *unconscious* is, by definition, largely unknown, presumed to be peopled by the blind, childish impulses and sub-human appetites or desires that we have repressed out of sight and bolted down under trapdoors, denizens that come out to play at night in our dreams.

As soon as we start discussing the unconscious mind we naturally resort to images, or vivid, graphic mental pictures drawn from everyday life that serve as counters for the indescribable: metaphors or similes pointing us to a reality that they can only partially disclose. Our language for communication with, or about, the unconscious mind, either within ourselves or from person to person, is the language of images, as artists and poets well understand. The closeness of images to the synthesizing meta-function is illustrated by the double meaning of *imagination*. It is both our mental faculty for forming images of external objects not present to the senses and the creative faculty of the mind.

These factors – the three meta-functions and the depth mind principle – taken together with the emotions – give us a model or theory of effective thinking. I include emotion or feeling, because emotion and motivation are closely related words. Emotion and motive energy are to the mind what electricity is to the brain. We need stimulus in order to be able to think. Some of it is self-generated; but some has to come from outside ourselves.

The part of the less conscious ranges of the mind in analysis is not often recognized, although, as we shall see later, the depth mind is capable of acting like a computer, unconsciously performing feats of analysis if it is correctly programmed and if certain conditions are present. People vary, however, in the capacity of their depth minds to work in this way.

EXERCISE 1: Using your depth mind
Try listing five ways in which you could increase your capacity to use your depth mind in a purposeful way:

1. ....
2. ....
3. ....
4. ....
5. ....

Repeat this exercise at the end of Part Two.

Intuition

Intuition, the apparent direct working of the mind without the intervention of any conscious reasoning process, may describe the instant and immediate eruption onto the surface of the mind of some *analysing* which has occurred at a subliminal level. For example, intuition may tell you that some situation exists for which you have no direct evidence. For instance, you sense that a competitor – despite their brave words – is close to bankruptcy. What may have happened is that your mind has registered a lot of information – imperceptible clues and hints and signs – through your eyes and ears which gets processed in your subconscious mind. Like a computer it then prints out your intuition – or a solution or decision.

Using your depth mind in business

I cannot explain this scientifically, but I was entirely convinced that, through the years, in my brain as in a computer, I had stored details of the problems themselves, the decisions reached and the results obtained; everything was neatly filed away there for future use. If the answer was not immediately apparent, I would let it go for a while, and it was as if it went the rounds of the brain cells looking for guidance that could be retrieved, for by next morning, when I examined the problem again, more often than not the solution came up right away. That judgement seemed to come to me almost unconsciously, and my conviction is that during the time I was not consciously considering the problem, my subconscious had been turning it over and relating it to my memory; it had been held up to the light of the experience I had had in past years, and the way through the difficulties became obvious. I am pretty sure other older men have had this same evidence of the brain's subconscious work.

This makes it all very easy, you may say. But, of course, it doesn't happen easily. That bank of experience from which I was able to draw in the later years was not easily funded.

Roy Thomson, *After I was Sixty* (Hamish Hamilton, 1975)

The discovery of the creative depth mind has, for some, upset the apple cart of received methods on how to teach decision-making in business schools, a field hitherto dominated by quasi-mathematic models of little relevance to the real world.

The depth mind or inner brain's role in our value thinking has yet to be explored more fully. Iris Murdoch once gave a good example in a radio interview, when she described going to a dinner party in Oxford with a friend during her student days. Among her fellow guests were two philosophers, who held forth as the evening went on. Going home, Iris Murdoch suddenly said to her friend, 'A [naming him] is a good man, and B [naming him] is a bad man.' Her inner brain had obviously reached those conclusions for her, and they suddenly surfaced into her consciousness.

Doubtless you will have had similar intuitions or surmises of what situations might be. So you can be sure that your inner brain is capable of analysing data that you may not have known you had taken in, and of comparing it with what is filed away in your memory bank. You will also have experienced the value thinking of the inner brain, if only in the form of guilt feelings, or even remorse, when it has made a moral evaluation or judgement of your own conduct. This unwanted and unasked contribution to your sanity is a reminder that the mind of the inner brain has a degree of autonomy from you. It is not your slave. Henry Thoreau once boldly suggested that 'the unconsciousness of man is the consciousness of God'.

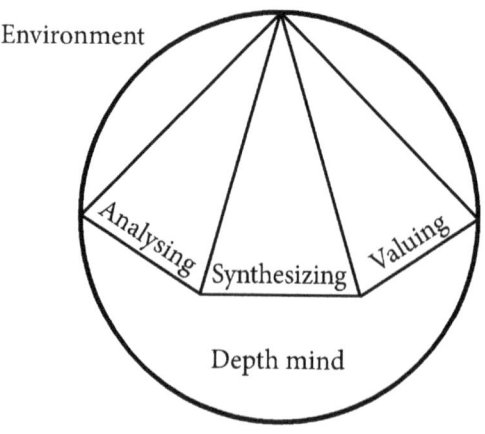

Meta-Functions of the Mind at Work

Sometimes, then, decisions and solutions, intuitions and ideas, seem to grow in the mind. It is as if there is a grain of sand in the oyster that grows into a pearl. It gives surprise and delight to a person fortunate enough to open it. Or it is like a seed that falls upon good ground, sends down roots and grows in a holistic way.

The word 'holistic', incidentally, comes from the Greek *holos*, meaning 'whole'. It was first coined in 1926 by the famous South African leader Jan Smuts, who, apart from being a soldier and statesman, was also a keen student of the agricultural sciences. He introduced it to describe the tendency of Nature to produce wholes from the ordered groupings of units. Babies are not parts that are assembled into a whole – they are wholes which *grow* larger. That's how Nature works. By analogy our minds work in the same way, some more than others. Women are inclined to be more holistic than men.

Holistic people tend to dislike too much analysis. Their pathway to understanding a person or a situation is to listen to the *story* of how they or it developed, from first beginnings to present state. The idea of *growth* appeals to them. For example, 'growing' a business – not just in size but in quality – may appeal to a holistic leader or entrepreneur far more than making a lot of money.

Holistic people have a head-start when it comes to creativity. By contrast, at the other extreme end of the spectrum, come the *analysts* – those who have developed that meta-function and equate it with thinking. Dissecting or taking things to bits, categorizing or classifying – useful though they can be – do not march so well under the creative banner. They have a part to play in the creative or innovative process but it's an ancillary one.

In the following case study, underline all the words or phrases that reflect a holistic bias or describe the creative activity of the depth mind.

CASE STUDY: C. S. FORESTER

As a novelist C. S. Forester is best known for his series of stories about Horatio Hornblower, a British naval officer in the era of the Napoleonic Wars. Forester once described his creative process thus:

> There are jellyfish that drift about in the ocean. They do nothing to seek out their daily food; chance carries them hither and thither, and chance brings them nourishment. Small living things come into contact with their tentacles, and are seized, devoured and digested. Think of me as a jellyfish, and the captured victims become the plots, the stories, the outlines, the motifs – use whatever term you may consider best to describe the framework of a novel. In the ocean there are much higher forms of life than the jellyfish, and every human being in the ocean of humanity has much the same experience as every other human being, but some human beings are jellyfish and some are sharks. The tiny little food particles, the minute suggestive experiences, are recognized and seized by the jellyfish writer and are employed by him for his own specialized use.
>
> We can go on with the analogy; once the captured victim is inside the jellyfish's stomach the digestive juices start pouring out and the material is transformed into a different protoplasm, without the jellyfish consciously doing anything about it until his existence ends with an abrupt change of analogy.
>
> In my own case it happens that, generally speaking, the initial stimulus is recognized for what it is. The casual phrase dropped by a friend in conversation, the paragraph in a book, the incident observed by the roadside, has some special quality, and is accorded a special welcome. But, having been welcomed, it is forgotten or at least ignored.

It sinks into the horrid depths of my subconscious like a waterlogged timber into the slime at the bottom of a harbour, where it lies alongside others which have preceded it. Then, periodically – but by no means systematically – it is hauled up for examination along with its fellows, and, sooner or later, some timber is found with barnacles growing on it. Some morning when I am shaving, some evening when I am wondering whether my dinner calls for white or red, the original immature idea reappears in my mind, and it has grown . . .

Sometimes I have been developing two different plots, both of them vaguely unsatisfactory, and then suddenly they have dovetailed together, like two separate halves of a jigsaw puzzle – the difficulties have vanished, the story is complete, and I am experiencing a special, intense pleasure, a glow of satisfaction – entirely undeserved – which is perhaps the greatest reward known to my profession . . .

C. S. Forester, *Long Before Forty* (Michael Joseph, 1967)

CHOOSE WORK THAT SUITS YOUR MIND

Do you think that the Canadian media mogul Roy Thomson could have written novels? Or that C. S. Forester could have used his depth mind, like Thomson did, as a kind of computer for decisions? No, for both men were led by their interests, temperament and aptitudes into fields where their energies could naturally engage the work most appropriate to them. Both men were fulfilled and happy in what they were doing; one creating a business empire and the other creating a world in fiction.

Now there are some individuals who are creative in more than one field of work, but they are the exceptions that prove the rule. So it's a good idea to check that you are operating in the one or two areas (some creative careers stem from combining modest talents in two different fields) where your natural abilities are highest. That is where you are most likely to make a creative contribution.

Don't label the fields with professional names such as 'Accountancy' or 'Medicine' at this stage. That will cut down your options. Keep a wide focus. Watch carefully, and see if any sparks jump – in your case – *between* the fields of intelligence on the table over the page, for that always suggests creative possibilities.

THE PARADOX OF CREATIVE THINKING

You may have seen a picture of Rodin's famous sculpture called *The Thinker*. The figure, carved out of marble, is bent almost double in thought with his hand supporting his head. To me it always suggests one part in the paradox of creative thinking – that it is a solitary business. On the one hand, you need unhurried time on your own to think, both about specific problems and about more general issues in your field. If you don't like your own company, your chances of making progress as a creative thinker are much reduced.

On the other hand, creative thinking is an intensely social activity. It can be hard to think without the stimulus and information inputs of other people's minds. This is one reason why people congregate in towns and cities. Our brains are open systems. Indeed, they are like radio transmitters and receivers, constantly communicating with other minds in an immense variety of ways, many of them barely recognized, let alone understood.

> Alex Graham, a hard-working chief executive of a medium-sized clothing business in Scotland, decided to take a sabbatical and sail around the world single-handed.

\multicolumn{2}{c}{**FIELDS OF INTELLIGENCE**}
Verbal	Good perception of relationships between words; ability to verbalize concepts or ideas in the chosen medium; good memory for written or spoken words. (Writers, poets, philosophers.)
Spatial	Good perception of spatial relationships; imaginative ability to think in three dimensions; good memory for spatial arrangement. (Artists, architects, some scientists, town planners.)
Numerate	Good perception of relationships between figures; ability to work out complex mathematical calculations mentally; good memory for numbers. (Pure mathematicians, financiers, accountants, some scientists, actuaries.)
Colour	Good perception of relationships between colours; ability to swiftly visualize colour schemes; good memory for colours. (Artists, dress designers, interior decorators.)
Musical	Good perception of relationships between sounds; ability to compose music mentally; good memory for tunes. (Composers, singers, jazz musicians.)
Mechanical	Good perception of relationships between parts of engines or similar systems; ability to invent new machines or systems, or adapt old ones to new purposes. (Engineers, mechanics, computer systems people.)
Organic	Good perception of relationships between parts and wholes in plants, animals or humans. (Doctors, biologists, environmentalists.)
People	Good perception of relationships between people; ability to create teams and organizations; an ability to help individuals develop.

After two weeks his yacht struck a rock and sank, but he struggled ashore on an uninhabited island out of sight of the Venezuelan coast.

'This is an opportunity,' he told himself. 'I have always wanted more time to think.' He had plenty of time – he was not rescued for three weeks – but he came up with no new ideas.

Why do you think that was?

A hermit or recluse may grow spiritually in solitary dialogue with God. But he or she is unlikely to become a clear or creative thinker. Thoughts are clarified in dialogue with others. The edges of our minds are sharpened on the whetstones of other brains. To think creatively we need a constant supply of stimulating social contact to stir, rouse or activate our ten billion brain cells. We build or feed upon the contributions of others, and they in turn feed upon us.

The greatest minds are often the most aware of how much they owe to intellectual intercourse with others – alive or dead – in their fields and in the wider areas of human knowledge and experience. We are pygmies, they say, standing on the shoulders of giants.

Bearing this paradox in mind, as a creative thinker you have to manage the balance between the necessary solitude or time on your own to think and the necessary social intercourse with others. This can take many forms, such as:

- attending a conference
- networking on the telephone
- reading a book
- visiting a library
- a conversation over dinner
- a meeting at work
- watching television
- listening to the radio

Your need here is for equilibrium: a balance between social interaction on the one hand and being alone on the other. If you feel as if you are lacking in the stimulation that comes from contact with other minds, you should maintain the balance by actively seeking the stimulation yourself – join a book club, arrange a dinner with friends or take part in a conference. Or if your social life is pressing, make the effort to create some time and space when you can be alone with your thoughts, by, say, getting up earlier in the morning. For in order to think creatively, you must listen to the faint signals from your depth mind and respond accordingly – and that requires time on your own.

KEY POINTS: HOW THE MIND WORKS

- Behind the more or less specific activities of problem solving, decision making and creative thinking, lies the more general stream of applied thinking. In its deliberate sense, thinking takes three main forms: analysing, synthesizing and valuing.
- Below our conscious thinking in these respects lie the changing depths of our unconscious minds where the work of thinking is also done. We cannot, for example, allocate analysis solely to the conscious mind, nor synthesis entirely to the unconscious. Both these meta-functions, and also valuing, can and do take place in the unconscious depth mind. Perhaps the first step towards improving your everyday thinking is to become more aware that purposeful thinking can take place on different levels of the mind.
- When it comes to any form of practical thinking it's important to choose the field where your mind works naturally at its best. We all have different mental talents. It is a rare person who is outstandingly creative in more than one field, although the creative orientation – refracted in the seven habits described in Part Two – may well influence the whole of one's life. As the author Henry Miller said, 'When the creative urge seizes one – at least, such is my experience – one becomes creative in all directions at once.'

CHECKLIST:
HOW YOUR MIND WORKS

Which of the three meta-functions of your mind do you think is the most – and least – developed?

	Most	Least
Analysing	☐	☐
Synthesizing	☐	☐
Valuing	☐	☐

Now answer Yes or No to the following:

	Yes	No
Have you ever experienced your depth mind working to bring together or synthesize two apparently different courses of action?	☐	☐
Have you ever left an issue requiring a decision for a period of time and then found that your mind has made itself up in favour of one option?	☐	☐
Would you describe your mind as holistic, in that you prefer to see the whole and dislike analysing or taking things to pieces?	☐	☐
Do you tend to make decisions about people rather slowly and relying upon your intuition?	☐	☐
Has your depth mind in the guise of conscience ever evaluated anything you have done or said some time after the event?	☐	☐
Are you clear in which field your natural interests and strengths as a thinker (see the Fields of Intelligence table) lie?	☐	☐
Do you have strengths in more than one field?	☐	☐
Do you balance time for thinking on your own with stimulating interaction with others?	☐	☐

PART TWO

SEVEN HABITS OF SUCCESSFUL CREATIVE THINKERS

Basically, habits are ways of acting that have become fixed through repetition. So much so, that we use the word 'habit' to imply the act of doing something unconsciously or without thinking about it in advance.

We all develop habits of thinking as well as of behaviour – some good and some not so good. They are our settled dispositions or tendencies to approach problems in certain ways. With frequent use they become second nature. Together, these habits constitute what might be called our prevailing disposition or mental make-up.

The seven habits described in Part Two are some of the characteristics of the more creative or innovative thinkers. Of course, not all such thinkers exemplify all seven habits. But you do need a critical mass of them.

Together, they colour in most of the map of creative thinking techniques – but not all of it by any means, for some areas may always remain unmapped.

By the time you have completed reading and working on Part Two you should:

- Have a clear idea of the **seven habits** that creative thinkers employ in their quest for new ideas.
- Understand your own position at present in relation to each of these habits and **identify steps** towards making them more a natural part of your own creative processes.
- Explore some of the key **personal qualities** or **characteristics** that support this seven-headed, prevailing disposition of mind of creative thinkers.

HABIT ONE: GOING BEYOND THE NINE DOTS

In *Training for Decisions* (1969) I introduced a puzzle called the Nine Dots. It has become very popular since then and has appeared in many other books. But I want you to think of it here as more than just a mental puzzle. Consider it as a model for exploring how your own mind works. For you, it could be the key to opening the door of the first habit that characterizes effective creative thinkers.

All you have to do in order to solve the problem is to connect up the nine dots below by drawing four straight lines but without taking your pen or pencil off the paper. You should give yourself three minutes to solve the problem. If you know the answer already, you may like to give the problem to a friend or colleague and observe how they go about it. Or you may like to try and find a *second* solution – for there are at least two answers.

Nine Dots

The most elegant solution is given overleaf. Do not turn over until you have completed your three minutes' work on the problem.

If you haven't been able to solve the problem ask yourself why not. Is it because you have unconsciously imposed assumptions, constraints or rules onto the problem? Incidentally, why did you miss the clue given in the chapter title?

To solve the puzzle with four straight lines, you must challenge your assumption that the 'rules' meant you had to stay within the dots.

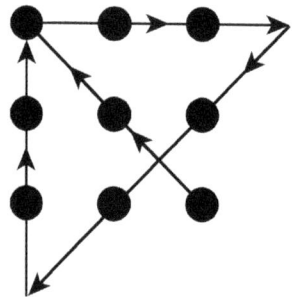

You can actually solve the puzzle with three straight lines. For you don't have to go through the centre of each dot. Even if the dots are drawn small that is still true, although the three lines may then extend a long way.

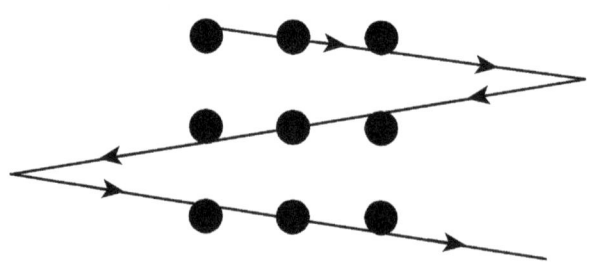

It's easy now, isn't it? Now try something a little more difficult.

Six Matches

Take six matches and place them on a flat surface. Arrange them into a pattern of four equilateral or equal-sided triangles. You must not break the matches. You have *five minutes* to find two solutions (there are more). Note that the triangles must be complete – no gaps at the corners!

Answers on page 206.

CHALLENGING ASSUMPTIONS

The first habit to develop as a result of your experience with the Nine Dots is to *challenge assumptions* – yours and other people's.

Do not misunderstand me. Assumptions, provided that they are made consciously, can play a vital part in creative thinking.

Einstein is famous for making such assumptions and thinking out their implications. 'Let me assume,' he said to himself, 'that I am riding on the back of a sunbeam, travelling through the universe with the speed of light. How would things look to me?' The eventual result was the General Theory of Relativity, through which Einstein led us to the knowledge that planets and stars move not because they are influenced by forces coming from other bodies in the universe, but because of the special nature of the world of space and time in the neighbourhood of matter. Light rays may travel in a straight line in the vast interstellar spaces, for example, but they are deflected or bent when they come within the field of influence of a star or other massive body.

The making of *conscious* assumptions like that one is a key tool in the tool-kit of a creative thinker. You are *deliberately* and *temporarily* making a supposition that something is true.

It is like making a move in a game of chess but still keeping your hand on the piece, so that you can replace it if you don't like the implications of the half-made move. 'No great discovery is made without a bold guess,' said Isaac Newton.

I have emphasized the words above in italics because this kind of exploratory thinking does need to be sharply distinguished from thinking based upon *unconscious* assumptions or preconceptions. We have all had the experience of taking something for granted as the basis for an opinion or action, and then subsequently finding that we had made an assumption – probably an unconscious one – that was unwarranted.

Watch out for these preconceptions! They are like hidden sandbanks outside the harbour mouth. Preconceived ideas are the ones you entertain prior to actual knowledge. The really dangerous ones are those that are completely below your level of awareness.

For we take on board all sorts of assumptions and preconceptions, often in the form of opinions or common sense, which on examination turn out to be unproven or debatable. They are the main impediments to having new ideas.

Received opinion on anything should be suspect. Once an idea is generally accepted it is time to consider rejecting it. But it can be very difficult to do that. For, to borrow Einstein's language, people in the mass can influence the space around them, deflecting the pure shaft of human thought.

'Few people', said Einstein, 'are capable of expressing with equanimity opinions which differ from the prejudices of their social environment. Most people are even incapable of forming such opinions.' We are social thinkers. Often great thinkers are rather solitary figures, possibly because they have a need to distance themselves psychologically from the powerful influences of received opinion.

When it comes to those dangerous unconscious assumptions, however, other people can be especially helpful to you.

They can sometimes alert you to the fact that you are *assuming* that something is the case without being aware that you are doing so. 'Why do you believe that?' they ask. 'What is your evidence? Who told you that you could not?'

Assumptive thinking is not the same as guessing. When we conjecture, surmise or guess we are really drawing influence from slight evidence. Guessing means hitting upon a conclusion, either wholly at random or from very uncertain evidence. Making an assumption is more like taking a tentative step. '*Supposing* we did it this way – how would it work? What would the consequences be?' It is not an answer – even a guessed answer – but it is a step that you can take if you are baffled, which might open up new possibilities.

It is most important to appreciate this difference between deliberately preconceived ideas and fixed ideas often unconsciously held. 'Preconceived ideas are like searchlights which illumine the path of an experimenter and serve him as a guide to interrogate nature,' said Louis Pasteur. 'They become a danger only if he transforms them into *fixed ideas* – that is why I should like to see these profound words inscribed on the threshold of all the temples of science: "The greatest derangement of the mind is to believe in something because one wishes it to be so."'

A reminder for managers
A thing is not right because we do it.
A method is not good because we use it.
Equipment is not the best because we own it.

Getting the balance right between imaginative thinking and critical thinking is essential for all creative thinkers, not least research scientists. Pasteur continued: 'Imagination is needed to give wings to thought at the beginning of experimental investigation into any given subject. When, however, the time has come to conclude, and to interpret the facts derived from observation, imagination must submit to the factual results of the experiments.'

Consequently, thinking will lead you to break or bend some of the rules that others consider set in stone. It is a fairly well-established rule in thinking that you should not base an argument on false premises. For the purposes of creative thinking, however, a 'false premise' in the shape of a bold and imaginative assumption may be just what you need in order to shatter your preconception. 'Daring ideas are like chessmen moved forward,' wrote Goethe. 'They may be beaten, but they may start a winning game.'

Thinking about thinking

Thus there may arise a scientific or critical phase of thinking, which is necessarily preceded by an uncritical phase. Our theories are our inventions; but they may be merely ill-reasoned guesses, bold conjectures, *hypotheses*. Out of these we create a world: not the real world, but our own nets in which we try to catch the real world.

If this was so, then what I originally regarded as the psychology of discovery had a basis in logic: there was no other way into the unknown, for logical reasons.

Karl Popper, *Unended Quest* (Routledge, 1992)

Managing assumptions is essential. It includes being able to both sense and challenge the *unconscious* constraints or rules that you or others impose on problems, situations, decisions, people, teams and organizations. And then to test or challenge them. They may, of course, withstand the questioning and you can then put your weight on them. Remember the Nine Dots!

Being able to manage assumptions also encompasses deploying them actively like scouts or pioneers of the main body of your mind. Here they act like tentative hypotheses. The more imaginative or far-fetched they are, the more likely they are to carry a payload of new possibilities or new ideas.

WIDENING YOUR SPAN OF RELEVANCE

The Nine Dots model is also important because it teaches us to look outside the framework and make connections beyond it. Once your pencil has connected with the invisible points *outside* the frame, the relation of the parts *inside* it falls into place, if you follow my reasoning. Creative thinkers tend to have a wide span of relevance. They habitually look beyond the Nine Dots for points in 'outer space' that connect to the problem and can suggest solutions.

Open this book out as far as you can. Stretch out your hand across the two pages. You may just be able to touch one edge with the tip of your thumb and the other with the tip of your little finger. If so, for you the open book is a 'span' wide. For a span is literally the distance from the end of the thumb to the end of a little finger of a spread hand. As you can see, it doesn't cover much ground – say 20 centimetres.

'Relevance' refers to what has bearing on the matter in hand. Therefore your span of relevance defines the field in which you look for ideas or material relating to your concerns. If your span of relevance is relatively narrow, you will look just around you or at least fairly close to home. The lines connecting what you regard as relevant to your mental focal point will be relatively short. You may fall into the Nine Dot trap.

The Case of the Blinkered Departmental Head

'Could I make a suggestion for improving the next leadership course for university heads of departments?'

'Yes, of course, that is what this evaluation session is all about. What have you in mind?' I replied.

'Well,' continued the chemistry departmental head, 'I found all your examples of leadership completely irrelevant. Universities are not armies or businesses or orchestras. Give us case studies and examples drawn from universities – British ones, that is, not American. You have to spell it out in terms of university management – make it more relevant.'

'Wouldn't that be rather too narrow a focus?' I began … 'And one other thing,' she added, 'please give us more examples of women as university leaders – in science departments.'

You can see the dilemma sparked by our conversation. There is a trade-off between narrow-focus/relevance and the more creative approach to thinking. What is relevant is a matter of *perception*. The narrow-span people tend to identify relevance if it is fairly obvious and at hand. Nothing will *connect* in their minds unless it is:

- clearly pertinent
- decisive in authority
- easily traceable
- immediately significant
- logically related

You can see these that criteria tend to confine you to your own professional or technical field. And they work extremely well for solving certain sorts of problems. A good doctor, for example, will be able to discern which of your symptoms or comments are relevant to his or her diagnosis and which fall outside the circle. But the generation of new ideas very often depends on making connections with points that are:

- not necessarily close to you personally in time or space
- perhaps completely outside your professional or technical sphere
- apparently irrelevant
- possibly disguised or buried – not easily accessible
- not in any logical or natural relationship
- lacking any authority

Of course, a more 'distant' connection is potentially more creative. For it is more likely to yield a 'new' idea in the sense that others won't have thought that way.

'Experience has shown,' wrote Edgar Allan Poe, 'and a true philosophy will always show, that a vast, perhaps the larger, portion of the truth, arises from the *seemingly irrelevant*.'

I have put *seemingly irrelevant* into italics to emphasize the point. To the person who encounters the Nine Dots puzzle for the first time, the space around the 'framed' dots is just that – seemingly irrelevant. In fact, it is relevant to solving the problem because it contains those two invisible points that your pencil must touch and turn upon.

In other words, what is seemingly irrelevant to one person may prove to be extremely relevant to another. It's largely a matter of perception. For example, musical instruments would appear to most farmers to be irrelevant to their problems, but not so Jethro Tull.

Farming in his native Berkshire in the early eighteenth century, the British agriculturist Jethro Tull developed a drill that enabled seeds to be sown mechanically, and so spaced that cultivation between the rows was possible in the growth period. Tull was an organist, and it was the principle of the organ that gave him his new idea. What he was doing, in effect, was transferring his knowledge of the technical means of achieving a practical purpose from one field to another.

In terms of the Nine Dots puzzle, notice that Tull goes well beyond the inner frame of the dots. He gazes over the mental fences around the industry of farming. He doesn't assume that the solution lies within the obvious span of relevance – other agricultural or industrial machinery – or within his immediate and present experience. Organs represent one of those points outside the Nine Dots with which he made his creative link or connection. Once made, then, it was a matter of strengthening and developing the resultant new line of thought.

FREEDOM FROM FIXED IDEAS

There is some competitive advantage in creative thinking when you are *not* too well-versed in the field in which you are trying to think new thoughts and to see what no one has seen before. The reason, of course, is that education or training in technical and professional knowledge inculcates a lot of assumptions. The subject rests on foundation assumptions and its superstructure is built on a framework of lesser ones. Because on the whole they work or deliver certain goods, they acquire authority and in the course of time become unquestioned. Rather than unlearning them it's easier if you haven't learnt them in the first place.

Therefore innovation often comes when a fresh mind,

untrammelled by these dead ideas and assumptions, enters a traditional industry. Sir Henry Bessemer, the British civil engineer who invented the Bessemer Process (1856) for converting molten pig iron into steel, once said: 'I had an immense advantage over many others dealing with the problem in as much as I had no fixed ideas derived from long-established practice to control and bias my mind, and did not suffer from the general belief that whatever is, is right.' But in his case, as with many 'outsiders', ignorance and freedom from established patterns of thought in one field were joined with knowledge and training in other fields.

Ask plenty of questions. *Why* are we doing it this way rather than any other? *What* are the criteria for success? *What* is the evidence that we are being successful? *When* did we last review these procedures? *Who* among our competitors is doing things differently and with *what* results? *Where* is the key research and development being done in this area?

These questions, repeated often, are like the points of a pneumatic drill digging up the hardened roads of organizational procedures. For you cannot sow seeds of change on tarmac roads. The practices and procedures of organizations are rather like roads. 'When a road is once built,' wrote Robert Louis Stevenson, 'it is a strange thing how it collects traffic, how every year as it goes on, more and more people are found to walk thereon, and others are raised up to repair and perpetuate it, and keep it alive.'

Lack of expert or specialized knowledge in a given field is no bar to being able to make a creative contribution. Indeed, too much knowledge may be a disadvantage. As Benjamin Disraeli said, we must 'learn to unlearn'.

James Dyson is the inventor of, amongst other things, the ubiquitous Ballbarrow, a dumper-truck-style plastic hull suspended on a large football. Another of his inventions is the

QUIZ
Write down the main occupation of the inventors of the following products:

INVENTION	OCCUPATION
1. Ballpoint pen	_____
2. Safety razor	_____
3. Kodachrome films	_____
4. Automatic telephone	_____
5. Parking meter	_____
6. Pneumatic tyre	_____
7. Long-playing record	_____

Answers on page 207.

Dual Cyclone, a vacuum cleaner which doesn't need a bag but filters the dirt by the centrifugal force of two cyclones. A multi-million-pound-a-year business has developed on the basis of his quality engineering products. Yet this outstanding innovator studied Classics at school and Design at the Royal College of Art. Dyson began with furniture design and then, led by his interest in technology, veered towards product design.

But how can someone trained as an artist make the leap to technological innovation? He says: 'Engineering is just a state of mind. You don't need a vast amount of knowledge. But I cling on to the belief that anyone can become an expert in a specific area in about six months, whether it's hydrodynamics for boats, cyclonic systems for vacuum cleaners, or wheelchair propulsion. I steer clear of projects that involve too much maths and try to stick to empirical things, ideas that require an Edisonian approach.'

The case of Francis Crick

The first success in the Laboratory of Molecular Biology came in 1953, when James Watson and Francis Crick discovered the double helical structure of DNA, which explains the nature of genetic information and the way that information is copied and passed on from parent to progeny. It was a discovery that transformed the face of biology. In 1945, when he was almost thirty, Crick had taken stock of his qualifications:

'A not-very-good degree, redeemed somewhat by my achievements at the Admiralty. A knowledge of certain restricted parts of magnetism and hydrodynamics, neither of them subjects for which I felt the least bit of enthusiasm. No published papers at all ... Only gradually did I realize that *this lack of qualification could be an advantage.* By the time most scientists have reached the age of thirty they are trapped in their own expertise. They have invested so much in one particular field that it is often extremely difficult, at that time in their careers, to make a radical change. I, on the other hand, knew nothing, except for a basic training in somewhat old-fashioned physics and mathematics and an ability to turn my hand to new things.'

What Mad Pursuit: A Personal View of Scientific Discovery
(Penguin, 1990)

What really matters, then, is knowing how to think. Creative thinking is essentially a set of habits or qualities. It doesn't reside in technical or professional knowledge, despite the fact that these play a part in both the initial thinking and, later, in bringing the idea to market in a technically feasible and commercially viable form. So, remember, your lack of qualifications or book knowledge could actually work to your advantage.

Barnes Wallis, who invented the bouncing bomb in the Second World War and subsequently helped to develop the Concorde supersonic airliner and the swing-wing aircraft, failed his London Matriculation Examination at the age of sixteen. 'I knew nothing,' he said in a television interview, 'except how to think, how to grapple with a problem and then go on grappling with it until you had solved it.'

DEFINING THE PROBLEM CORRECTLY

As the concept of lateral thinking suggests, a 'problem is a solution in disguise'. If the problem is incorrectly defined or analysed, then it's almost impossible to recognize a solution.

The object of analysis is, therefore, clarity of thought, for clear thinking should precede and accompany creative thinking. Ask yourself, What is the focus of your thinking? Is it some necessity, some everyday problem, or a resource that could be exploited in several different ways? If it is a problem, what are the success criteria for any satisfactory solution?

Check your definition of the problem. Are you rating symptoms rather than the disease? There are often several equally valid (but not equally obvious) ways of defining any problem. But each definition is a general statement of a potential solution to the problem. So different definitions are worth collecting: they are signposts for different avenues of thought. The definition you settle upon may have a powerful influence in programming your depth mind. If it leads nowhere, try another definition.

Edward Jenner's discovery of vaccination illustrates how useful it can be to be able to redefine the problem. At the end of the eighteenth century, Jenner took the first step towards ending the scourge of smallpox when he turned from the question of why people caught the disease to why dairymaids did not: the answer being that they were immunized by exposure to the relatively harmless cow-pox.

> Two men were walking in the African bush when they met a very hungry cheetah who eyed them ferociously. One of the men fished out some running shoes from his knapsack and bent down to put them on. 'Why are you doing that?' cried his companion in despair. 'Don't you know that cheetahs can run at over sixty miles per hour?' 'Yes, yes,' the first replied as he finished tying the laces. 'But I only have to outrun you.'

The best advice is not to focus too strongly on any aspect of the problem. You should, rather, learn to think generally about it, like a scientist scanning a problem area for clues. Let it speak to you. 'Whatever the ultimate object of his work,' wrote Hazel Rossotti, in *Introducing Chemistry* (Penguin, 1978), 'the experimental chemist's immediate aim is to ask suitable questions of the sensible bodies he is studying and to let them answer for themselves. It is the chemist's job to observe and report the answers with minimal distortion; only then can he attempt to interpret them'. These attitudes, a proper detachment and objectivity, are relevant to creative thinkers in the conscious phases of their work.

It is so easy to introduce subjective elements – such as those troublesome unconscious assumptions or constraints – into the problem or matter under review. Patient analysis and restructuring of the parts, taking up different perspective points in your imagination from which to view it: all these will deepen your understanding of the problem if they do not fairly soon release within you, like a cash dispenser, the right solution or at least tell you the right direction in which to advance.

The following sets of questions are designed to help you to solve any kind of problem. Try applying them to one or two problems that confront you now.

Understanding the Problem

1. Have you defined the problem or objective in your own words?
2. Are there any other possible definitions of it worth considering? What general solutions do they suggest?
3. Decide what you are trying to do. Where are you now and where do you want to get to?
4. Identify the important facts and factors. Do you need to spend more time on obtaining more information? What are the relevant policies, rules or procedures?
5. Have you reduced the complex problem to its simplest terms without over-simplifying it?

Solving the Problem

6. Have you checked all your main assumptions?
7. Ask yourself and others plenty of questions. What? Why? How? When? Where? Who?
8. List the obstacles that seem to block your path to a solution.
9. Work backwards. Imagine for yourself the end state, and then work from there to where you are now.
10. List all possible solutions, ways forward or courses of action.
11. Decide upon the criteria by which they must be evaluated.

12. Narrow down the list to the *feasible* solutions, that is, the ones that are possible given the resources available.
13. Select the optimum one, possibly in combination with parts of others.
14. Work out an implementation programme complete with dates or times for completion.

Evaluating and Implementing the Decision

15. Be sure that you have used all the important information.
16. Check your proposed solution from all angles.
17. Ensure that the plan is realistic.
18. Review the solution or decision in the light of experience.

Now you know all the theory, see if you can solve these problems:

1. Who Owns the Zebra?
There are five houses, each with a front door of a different colour, and inhabited by men of different nationalities, with different pets and drinks. Each man eats a different kind of ice cream.
   The Englishman lives in the house with the red door.
   The Spaniard owns the dog.
   Coffee is drunk in the house with the green door.
   The Ukrainian drinks tea.
   The house with the green door is immediately to the right (*your* right) of the house with the ivory door.
   The chocolate ice cream eater owns snails.
   Vanilla ice cream is eaten in the house with the yellow door.
   Milk is drunk in the middle house.

The Norwegian lives in the first house on the left.
The man who eats strawberry ice cream lives in the house next to the man with the fox.
Vanilla ice cream is eaten in the house next to the house where the horse is kept.
The raspberry ice cream eater also drinks orange juice.
The Japanese eats banana ice cream.
The Norwegian lives next to the house with the blue door.
Now, who drinks water? And who owns the zebra?

2. The Swimming Pool
A man wants to double the size of the square swimming pool in his garden. There is a tree growing at each corner, like this:

How does he double the size of the pool, still keeping it square, and without cutting down any of the trees?

3. The Restaurant Meal
Three women each have two daughters, and they all go into a restaurant for a meal. There are only seven vacant seats in the restaurant, but each has a seat to herself. How do they manage it?

4. Relatives
A doctor in London has a brother in Manchester who is a lawyer. But the lawyer in Manchester does not have a brother in London who is a doctor. Why?

5. Bottled Money
If you put a small coin piece in an empty bottle and replace the cork, how can you get the coin out of the bottle without taking out the cork or breaking the bottle?

6. Farmer's Choice
A farmer has twelve sheep and decides to put each of them in a separate pen; but to make the pens he has only twelve long hurdles and six half-length hurdles. How does he do it?

7. Prisoner's Escape
Harry is languishing in jail in Mexico. The cell has multiple locks on the door; the walls are made of concrete extending two storeys into the ground; the floor is made of packed earth. In the middle of the ceiling, eight feet above Harry, is a skylight just wide enough for him to squeeze through. The cell is bare.

One night, in desperation, he has an idea. He starts digging in the floor, knowing he can never tunnel out.

What is his plan?

8. Drinking Glasses
Six drinking glasses stand in a row, with the first three full of water and the next three empty. By handling and moving only one glass, how can you arrange the six glasses so that no full glass stands next to another empty one?

9. The Bicycles and the Fly
Two boys on bicycles, twenty miles apart, begin racing straight towards each other. The instant they start, a fly on the handlebar of one bicycle starts flying straight towards the other cyclist. As soon as it reaches the other bicycle, it turns and starts back. The fly flies back and forth in this way, from handlebar to handlebar, until the bicycles meet.

If each bicycle has a constant speed of ten miles an hour, and the fly flies at a constant speed of fifteen miles an hour, how far does the fly fly?

10. The Three Ties

Mr Brown, Mr Green and Mr Black are lunching together. One of them is wearing a brown tie, one a green tie, one a black.

'Have you noticed,' says the man with the green tie, 'that although our ties have colours that match our names, not one of us is wearing a tie that matches his *own* name?'

'By golly, you're right!' exclaims Mr Brown.

What colour tie is each man wearing?

Answers on page 207–9.

KEY POINTS: GOING BEYOND THE NINE DOTS

- We tend to impose invisible frameworks around problems or situations, whereas the solution may begin at some point outside that mental box or boundary. Therefore learn to think beyond the Nine Dots.
- Develop your awareness of the jungle of tangled misconceptions, preconceptions and unconscious assumptions within you. Welcome others when they challenge or test your assumptions.
- The ability to explore possible ways forward by making some deliberate assumptions is important. They are to be made without commitment, like trying on new clothes in a shop before buying (or not buying) them.
- Today's commonsense assumptions are very different from those of fifty years ago. What will common sense be like in fifty years' time?

- People with a narrow span of relevance are thinking within the tramlines and boundaries of their own experience or field of work. Leap over the wall! Develop a wide span of relevance, for there are connections between every other industry in the world and your one – if only you could see them.
- 'He is most original who adapts from the most sources.' You will be creative when you start seeing or making connections between ideas that appear to others to be far apart: the wider the apparent distance the greater the degree of creative thinking involved.

> *Discovery consists of seeing what everyone has seen and thinking what nobody has thought.*
> Anon

CHECKLIST:
GOING BEYOND THE NINE DOTS

Answer Yes or No to the following:	Yes	No
If you didn't know the answer to the Nine Dots problem, were you able to solve it in three minutes?	☐	☐
Do you appreciate the difference between such unconscious assumptions and the deliberate if tentative ones that all thinkers must make – 'supposing...' 'if...'?	☐	☐
Do you find that you are now more aware of the unexamined assumptions your colleagues tend to make?	☐	☐
Can you balance imaginative flights of thought with realistic assessment based on fact?	☐	☐
Do you have a tendency to dismiss ideas, people or things as irrelevant?	☐	☐
Can you think of anyone you know personally who has a much wider span of relevance than your own?	☐	☐
Do you sometimes blame your lack of success as a thinker on not going to university or not having specialist knowledge in a particular field?	☐	☐

HABIT TWO: WELCOMING CHANCE INTRUSIONS

As I was settling down to write this chapter I heard a shout at the back door. 'Hello, anyone at home?' It's the moment that any manager dreads – *another* interruption. Up you get to greet the unexpected visitor, leaving the links in the chain of thought you were constructing scattered over the table.

Yet there is a great deal of evidence to support the view that often such chance events can provide the missing link in the creative process. The second habit of creative thinkers is to be especially aware and observant of everything that is happening in the vicinity while they are thinking or working on a problem, which could of course be over a period of months or years. That radar-like awareness takes in *chance* happenings in the environment.

CHANCE – THE INVENTOR'S LUCK

It is interesting to reflect how many inventions have been the result of such unexpected or chance occurrences as befell Sir Alastair Pilkington at his kitchen sink. You may recall that before the development of the float process by a research team led by Sir Alastair, glass-making was labour-intensive and time-consuming, mainly because of the need for grinding and polishing surfaces to get a brilliant finish.

Pilkington's proprietary process eliminates this final manufacturing stage by floating the glass, after it is cast from a melting furnace, over a bath of molten tin about the size of a tennis court. The idea for 'rinsing' glass in this way came to Sir Alastair when he stood at his kitchen sink washing dishes. The float process gives a distortion-free glass of uniform quality with bright, fire-polished surfaces. Savings in costs are considerable. A float line needs only half the number of workers to produce three times as much glass as old production methods. Since the introduction of the process in 1959, it is estimated to have earned Pilkington over $600 million in royalties.

The classic example of the part that chance plays in invention is the discovery of penicillin by Sir Alexander Fleming. But there are plenty of others:

Saccharine

The sweetening effect of saccharine was accidentally discovered by a chemist who happened to notice a sweet taste on his hands after working with certain chemicals.

Offset printing

Ira W. Rubel observed the effects when a feeder failed to place a sheet of paper in a lithograph machine, and the work on the printing surface left its full impression upon the printing cylinder: it led him to invent the offset method of printing.

Mirror galvometer

The idea of the mirror galvometer first occurred to William Thompson when he happened to notice a reflection of light from his monocle.

Vulcanization

Charles Goodyear discovered the vulcanization of rubber in 1839 by similar observation of a chance event. He had been experimenting for many years to find a process of treating crude or synthetic rubber chemically to give it such useful properties as strength and stability, but without success. One day, as he was mixing rubber with sulphur, he spilt some of the mixture onto the top of a hot stove. The heat vulcanized it at once. Goodyear immediately saw the solution to the problem that had baffled him for years. As he pointed out, however, chance was by no means the only factor in his useful discovery:

> I was for many years seeking to accomplish this object, and *allowing nothing to escape my notice* that related to it. Like the falling apple before Newton's gaze, it was suggestive of an important fact to one whose mind was previously prepared to draw an inference from *any occurrence that might favour the object of his research*. While I admit that these discoveries of mine were not the result of scientific chemical investigation, I am not willing to admit that they are the result of what is commonly called accident. I claim them to be the result of *the closest application and observation*.

I have put some of Goodyear's words into italics because they highlight the importance of having a wide focus of attention and keen powers of observation. His message is admirably summed up in Pasteur's famous words: 'In the field of observation, chance favours only the prepared mind.'

What does it mean for you to have a prepared mind? You have to be purposeful in that you are seeking an answer or solution to some problem. You have to become exceptionally sensitive to any occurrence that might be relevant to that search. You must have the experience to recognize and interpret a clue when you see or hear one. That entails the ability to remain alert and sensitive for the unexpected, while watching for the expected. You will have to be willing to invest a good deal of time in fruitless work, for opportunities in the form of significant clues do not come often. In those long hours, experiment with new procedures. Expose yourself to the maximum extent to the possibility of encountering a fortunate accident.

Perhaps in good management there is no such thing as luck as most people conceive it. Luck is opportunity meeting preparation.

PRACTISE SERENDIPITY

Serendipity is a happy word. Horace Walpole coined it to denote the faculty of making lucky and unexpected 'finds', by accident. In a letter to a friend (28 January 1754) he says that he coined it from the title of a fairy story *The Three Princes of Serendip* (an ancient name for Sri Lanka), for the princes 'were always making discoveries, by accidents and sagacity, of things they were not in quest of'.

If serendipity suggests chance – the finding of things of value when we are not actually looking for them – the finder must at least be able to see the creative possibilities of his or her own discovery. Edison was seeking something else when he came across the idea of the mimeograph. But he had the good sense to realize that he had made an important discovery and soon found a use for it.

Serendipity goes against the grain of narrow-focus thinking, where you concentrate your mind on a single objective or goal to the exclusion of all else. It invites you to have a wide span of attention, wide enough to notice something of significance even though it is apparently irrelevant or useless to you at present.

The three princes in the story were travellers. Explorers into the unknown often make unexpected discoveries. As the proverbial schoolboy knows, Christopher Columbus was seeking a new sea route to Asia when he discovered the New World. He thought he had reached India, which is why he called the natives he found there Indians. When you travel you should do so in a serendipitous frame of mind. Expect the unexpected. You may not discover America but you will have some happy and unexpected 'finds'.

'Thinking will always give you a reward, though not always what you expected.' These wise words were spoken by the Canadian entrepreneur and businessman, Lord Thomson of Fleet.

When you are thinking you are travelling mentally, you are on a journey. For genuine thinking is always a process possessing direction. Look out for the unexpected thoughts, however lightly they stir in your mind. Sometimes an unsuspected path or by-way of thought that opens up might be more rewarding than following the fixed route you had set yourself.

KEY POINTS: WELCOMING CHANCE INTRUSIONS

- Serendipity means finding valuable and agreeable things when you are not seeking them.
- You are more likely to be serendipitous if you have a wide attention span and a broad range of interests.
- Developing your capacity for creative thinking will bring you rewards, but they may not be the ones you expect now.
- The transfer of technology from one field to another, usually with some degree of alteration and adaptation, is one way in which you can make a creative contribution.
- You may be familiar with a body of knowledge or technical capability unknown to others in your field because you have worked in more than one industry. Or it may come about as a result of your travels to other countries.
- Things that happen unpredictably, without discernible human intention or observable cause, can be stitched into the process of creative thinking.
- To see and recognize a clue in such unexpected events demands sensitivity and observation.
- To interpret the clue and realize its possible significance requires knowledge without preconceptions, imaginative thinking, the habit of reflecting on unexplained observations – and some original flair.

Chance favours only the prepared mind.
Louis Pasteur

CHECKLIST:
WELCOMING CHANCE INTRUSIONS

	Yes	No
'Chance favours the prepared mind.' Can you think of an opportunity that you were able to seize because you were ready for it?	☐	☐
Has a chance event or meeting played a vital role in any creative work that you have undertaken?	☐	☐
Are you aware of the amazing number of inventions or discoveries in which chance events within the background turned out to be the breakthrough factors?	☐	☐
Have you developed the habit of noticing any occurrence on the grounds that it might have some relevance to the problems you are working on?	☐	☐
Have you ever made any discoveries or stumbled across opportunities when you weren't looking for them?	☐	☐
Did you meet your husband/wife/partner by chance?	☐	☐
Would you subscribe to the view that everything is connected with everything else, though we see only very few of the connections?	☐	☐

5

HABIT THREE: LISTENING TO YOUR DEPTH MIND

A remarkable fact about hearing is that you can sometimes hear things happen *inside* your head, such as blood pulsating in arteries. Using that analogy, creative thinkers develop a habit of listening to what is going on inside their depth or subconscious minds. Turn back to page 30 and re-read C. S. Forester's description of how his mind works and you will see what I mean.

The versatility of the depth mind in gifted creative thinkers is truly astonishing. It can duplicate and sustain those three abilities – analysing, valuing, synthesizing – at a deeper level and in a kaleidoscopic range of combinations. It can also work on four or five different 'projects' at once. As Tchaikovsky wrote:

> Sometimes I observe with curiosity that uninterrupted activity, which – independent of the subject of any conversation I may be carrying on – continues its course in that department of my brain which is devoted to music. Sometimes it takes a preparatory form – that is, the consideration of all details that concern the elaboration of some projected work; another time it may be an entirely new and independent musical idea . . .

HABIT THREE: LISTENING TO YOUR DEPTH MIND

We can all synthesize consciously. We can put two and two together to make four, or we can assemble bits of leather to make a shoe. But creative synthesis, as we have seen, is likely to be characterized by the combination of unlikely elements, distant from or apparently (to others) unrelated to one another. And the raw materials used will have undergone a significant transformation. The depth mind comes into its own when this kind of synthesis is required.

An obvious analogy is the womb, where after conception a baby is formed and grown from living matter. The word 'holistic', which applies to Nature's tendency to grow wholes from seeds, aptly applies to the synthesizing processes of the inner brain in the realm of ideas. Hence a new idea, concept or project is sometimes called a 'brain-child'. The great engineer Isambard Kingdom Brunel wrote about the Clifton Suspension Bridge in his diary as if it were a person: 'My child, my darling is actually going on – recommended last week – Glorious!'

The womb analogy is rather better than 'incubation', a term popularized by Graham Wallas in *The Art of Thought* (Harcourt, 1926). To incubate means to sit upon eggs so as to hatch them by the warmth of one's body. But the process is much more like a seed being implanted and fusing with that is another already present, which then grows by a form of accretion.

HOW TO BE MORE CREATIVE

Creative thinking cannot be forced. If you are working on a problem and getting nowhere, it is often best to leave it for a while and let your subconscious mind take over. Your mind does not work by the clock although it likes deadlines. Sometimes the answer will come to you in the middle of the night.

Grasping the principle of the depth mind – the subconscious and unconscious part of the brain – could open the door for you to this second key habit of mind. Many people are still not even aware that their depth minds can carry out important mental functions for them, such as synthesizing parts into new wholes or establishing new connections while they are engaged in other activities. Therefore they don't listen with their inner ears.

Imagine your mind is like a personal fax machine. It would be nice and tidy if you could sit down for an hour each morning before breakfast and receive inspired fax messages from your depth mind. But it isn't like that. The fax machine might start whirring at any time of the day or night.

According to Wallas, the first characteristic of original thinking, in a wide spectrum of fields, is a period of intense application, of immersion in a particular problem, question or issue (preparation). This is followed by a period when conscious attention is switched away from the topic, either by accident or design (the incubation phase). Sometimes a sudden flash of insight or intuition (illumination) will follow, then there is a period when the idea is subjected to critical tests and then modified (the verification stage). The stages are summarized opposite:

My own perspective is slightly different. As I see it, there is a conscious phase when you are aware of predominantly trying to analyse the matter that has engaged your attention. You may play around with some restructuring of it (synthesizing). Some valuing will enter into it – 'Is it worthwhile spending time on this project?' and the imagination may get to work, picturing some of the obvious solutions that occur to you, or their consequences. You may also be giving yourself advice or asking yourself questions such as, 'Remember not

THE CREATIVE PROCESS	
Preparation	Laying the groundwork. You have to collect and sort the relevant information, analyse the problem as thoroughly as you can, and explore possible solutions.
Incubation	This is the depth mind phase. Mental work – analysing, synthesizing and valuing – continues on the problem in your subconscious mind. The parts of the problem separate and new combinations occur. These may involve other ingredients stored away in your memory.
Illumination	The Aha! experience. A new idea emerges into your conscious mind, either gradually or suddenly like a fish flashing out of the water. These moments often occur when you are not thinking about the problem but are in a relaxed frame of mind.
Verification	This is where your valuing faculty comes into play. A new idea, insight, intuition, hunch or solution needs to be thoroughly tested. This is especially so if it is to form the basis for action of any kind.

to accept the first solution that comes to mind,' or, 'Am I making any unconscious assumptions?' This phase corresponds to Wallas's 'preparation' phase, though the label is misleading because we revert quite often to this conscious working of our minds in ordinary circumstances anyway.

When we are not engaged, these activities of analysing, synthesizing and valuing can continue – but they do not do so invariably – at the level of our depth or unconscious minds. We may then receive the products of such subliminal thinking in a variety of ways. The American poet Amy Lowell, for instance, says, 'I meet them where they touch consciousness, and that is already a considerable distance along the road to evolution.'

Far from this reception of an idea from the unconscious mind with the consciousness being the end of the story, it is only a halfway stage. During the process of working out, other fresh ideas and developments of a creative kind will still occur.

Solvitur ambulando – Solve it while walking

'There is an old saying "well begun is half done",' said the poet John Keats to a friend who inquired if his ideas came to him fully-formed. ''Tis a bad one. I would use instead, "Not begun at all until half done."'

There are some cases, indeed, where an idea or concept appears initially in its finished and fully-fledged form, but they are the exceptions. What is usually received is less than that. You have to work it out. In the process of doing this the idea may be developed, adapted or changed, and new ideas or materials will be added to the melting pot. As Sir Huw Wheldon, the television producer, once said in a televised lecture, 'Programmes are made in the making.'

This approach may sound rather untidy, even chaotic. And so it is. It goes against the grain for those who have been indoctrinated to seek finished ideas before going to work. But it adds greatly to the interest and excitement of work if you don't know what is coming next. 'I have never started a poem yet whose end I knew,' said the American poet Robert Frost. Creative thinking has to be an adventure.

Knowing when to stop thinking and to try working out

an idea is an important act of judgement. If you are premature you will waste a lot of time fruitlessly chasing ideas that are not right. But if you have a working clue don't wait too long! John Hunter, the famous eighteenth-century British surgeon and physiologist, had considerable influence as a teacher. His most brilliant pupil was Edward Jenner, who had already begun to think that he could prevent smallpox by vaccination. 'Don't think,' Hunter advised. 'Try it! Be patient, be accurate!' And the pupil spent many years in painstaking observation. In due course, Jenner discovered the smallpox vaccine.

Creative thinking leads you to the new idea; innovation involves actually bringing it into existence. To give something form – to bring an idea actually into existence – requires a range of skills and knowledge beyond the more cerebral ones we have been considering in this book so far. The artist is an obvious case in point. Leonardo da Vinci may have lain in bed in his darkened chamber going over and over again in his imagination his observations of the previous day and various ideas 'conceived by ingenious speculation'. But when he awoke the next morning and went into his studio he had the skill to make models, draw and paint with a consummate craftsmanship acquired over a lifetime. He may not have translated all his original ideas into existence – in the cases of the helicopter and submarine, the technology was lacking – but he could certainly express them in detailed drawings.

If you are using prime time for thinking along a certain line and nothing happens, stop. Instead of investing more time – throwing good money after bad – analyse the problem again and see if you can come up with a new approach. Usually your frustration will be caused by one of the following mental road blocks:

Lack of a starting point	Possibly the problem or opportunity seems so large that you don't know where to start. If so, make a start anywhere. You can always change it later. Inspiration comes after you've started, not before.
Lack of perspective	Perhaps you are too close to the problem, especially if you have lived with it a long time or have been worrying about it incessantly. Try leaving it for a week. Simply explaining it to others may help. They may see new angles.
Lack of motivation	Do you want it to happen enough? Creative thinking requires perseverance in the face of surmountable difficulty. If you are too easily put off it may be a sign that deep down you lack the necessary motivation. Reinvigorate your sense of purpose.
Lack of consultation	Did you consult others? Remember, creative thinking is a social activity.

The processes of analysing a problem or identifying an objective are themselves means of programming the mind. Possible solutions and courses of action almost instantly begin to occur to us. Where there is a time delay, this means that the deeper parts of the brain have been summoned into action and have made what contribution they can.

Do you believe in the depth mind? By belief I don't mean that you are merely prepared to agree to the proposition that it exists, or even that it can work creatively for some people. By belief I mean something akin to personal faith. Do you *trust* your depth mind? Do you accept that it can, does and will work for you?

If your answer is an unreserved 'Yes' or a more cautious 'I hope so' or 'I would like to think so', some consequences for action follow. The fundamental principle is to work with Nature, not against it. You will see now in that natural context how important *preparation time* is for creative thinking: careful and clear analysis, conscious imagining or synthesizing (using such techniques as brainstorming and solo brainstorming), and exercising the valuing function of thought in a positive rather than negative way.

If you are planning to experiment and try a session before breakfast, it's always useful to have a preparation phase the night before. Imagine yourself as a house decorator, scraping down the woodwork and filling in holes and priming here and there, prior to painting a first coat next day.

SLEEPING ON PROBLEMS – AND SOLUTIONS

When you are relaxed in bed before going to sleep, it is good to think about an issue requiring creative thought of the unconscious mind in the inner brain. The value of doing so has been long known. As Leonardo da Vinci wrote, 'It is no small benefit on finding oneself in bed in the dark to go over again in the imagination the main lines of the forms previously studied, or other noteworthy things conceived by ingenious speculation.'

Of course, you might actually dream of a solution. Why we dream is still largely a mystery. Dreams are extraordinary creations of our imagining faculty in the realm of mind. Sometimes they have messages from the hidden parts of our brain for us, not by fax this time but coded in an alien language of images. Some people in history have had a gift for interpreting dreams, not always in ways that endeared them to others. 'Here comes that dreamer!' Joseph's older brothers said to one another. 'Come now, let's kill him ... Then we'll see what comes of his dreams.' (Genesis 37:19ff)

You may like to try the experiment of jotting down fragments of any dreams you can recall when you wake up. See how many suggestions or meanings you can discern in them. Even if they do not solve your problems, dreams may reveal your true feelings and desires, especially if these have been suppressed for too long. As the novelist William Golding says, 'Sleep is when all the unsorted stuff comes flying out as from a dustbin upset in a high wind.'

Occasionally you will be rewarded by a real clue in your dreams. In *Desert Island Discs* (Kimber, 1979) Roy Plomley talks of one such instance involving Sir Basil Spence, the distinguished architect who designed Coventry Cathedral:

> In designing a project of such vast size and complexity there were bound to be snags. He told me that at one point, when he was held up by a particular technical difficulty, he had an abscess on a tooth and went to his dentist, who proposed to remove the molar under a local anaesthetic. As soon as he had the injection, Spence passed out. During the short time he was unconscious he had a very vivid dream of walking through the completed cathedral, with the choir singing and the organ playing, and the sun shining through the stained glass windows *towards* the altar – and that is the way he subsequently planned it. Another inspiration was received when, flipping through the pages of a natural history magazine, he came across an enlargement of the eye of a fly, and that gave him the general lines for the vault.

Quite why sleep plays such an important part in helping or enabling the depth mind to analyse, synthesize and value is still a mystery. Dreams suggest an inner freedom to make all sorts of random connections between different constellations of brain cells. There may be some sort of shaking up of the kaleidoscope, resulting in new patterns forming in the mineshafts of the mind. We just do not know. This ignorance of *how* the brain works, in this respect, does not matter very much. What matters is that it *does* work. As the Chinese proverb says, 'It does not make any difference if the cat is black or white as long as it catches mice.'

There is an element of mystery about this creative work that can go on in our sleep. Robert Louis Stevenson spoke of 'those little people, my brownies, who do one half of my work for me while I am fast asleep, and in all human likelihood do the rest for me as well, when I am wide awake and fondly suppose I do for myself'.

KEY POINTS: LISTENING TO YOUR DEPTH MIND

- According to an old English proverb, 'There is a great deal of unmapped country within us.' In part, creative thinking is exploring an unknown hinterland: the myriad galaxies of your inner brain.
- The meta-functions of the mind at work – analysing, valuing and synthesizing – resonate in the whole mind. Your depth mind – your subsequent and unconscious mind – replicates these functions. It can, for example, dissect something for you, just as your stomach juices can break down food into its elements.
- Your depth mind is capable of more than analysis. It is the seat of your memory, or at least a part of it, and it's

also the repository of values that lie too deep for words. It is also a workshop where creative syntheses can be made by an invisible workmanship.
- You have most probably experienced the beneficial effects of sleeping on a problem, and waking up to find that your mind has made itself up. Use that principle by programming your depth mind for a few minutes as you lie in the dark before you go to sleep.
- Your dreams may occasionally be directly relevant. It is much more likely, however, that some indication, clue or idea will occur to you after 'sleeping on it'. Perhaps during your waking hours, or, for instance while you are shaving or washing the dishes, the idea will dart into your mind.

While the fisher sleeps, the net takes the fish.
Ancient Greek Proverb

CHECKLIST:
LISTENING TO YOUR DEPTH MIND

	Yes	No
Answer Yes or No to the following:		
Do you have a friendly and positive attitude to your inner brain. Do you *expect* it to work for you?	☐	☐
Where possible, do you build into your plans time to 'sleep on it', so as to give your inner brain an opportunity to contribute?	☐	☐
Do you deliberately seek to employ your depth mind to help you to:		
– analyse a complex situation?	☐	☐
– restructure a problem?	☐	☐
– reach value judgements?	☐	☐
Have you experienced waking up next morning and finding that your unconscious mind has resolved some problem or made some decision for you?	☐	☐
Do you see your inner brain as being like a computer? Remember the computer proverb: 'Rubbish in; rubbish out.'	☐	☐
Do you keep a notebook or pocket tape recorder at hand to capture fleeting or half-formed ideas?	☐	☐
Do you think you can benefit from understanding how the depth minds of other people work?	☐	☐

HABIT FOUR: SUSPENDING JUDGEMENT

Exhilaration is that feeling you get just after a bright idea hits you, and just before you realize what's wrong with it. Postpone that sudden departure of excitement through evaluation. Give the new idea some room to breathe. As your excitement continues, it may reveal new possibilities.

Creative people develop the habit of being able to separate the valuing meta-function of mind from the other meta-functions of synthesizing, imagining and holistic thinking. If you want to encourage new ideas don't evaluate too soon: give your seeds a chance to grow.

'In the case of the creative mind,' the German poet Johann Schiller wrote some two hundred years ago, 'it seems to me it is as if the intellect has withdrawn its guards from the gates. Ideas rush in pell mell and only then does it review and examine the multitude. You worthy critics, or whatever you may call yourselves, are ashamed or afraid of the momentary and passing madness found in all real creators ... Hence your complaints of unfruitfulness – you *reject too soon and discriminate too severely.*'

DON'T CRITICIZE YOUR OWN IDEAS PREMATURELY

Take those guards of self-criticism away from the gates of your mind! Ideas should not need passports to come in and out of your mental house. For we do tend to post 'thought police' on our minds. We criticize or evaluate our own ideas – or half-ideas – far too soon. Criticism, especially the wholly negative kind, can be like a cold, white frost in spring; it kills off seeds and budding leaves. If we can relax our self-critical guard and let ideas come sauntering in, then we shall become more productive thinkers.

Don't let evaluation interfere with idea fluency. Keep ideas separate. Be as prolific as you can with them until you find one that satisfies you. Then try to translate it into the form you want.

BEWARE OF QUICK-FIRE CRITICS

Any sensible person should, of course, be open to the criticism of others. It is one of the duties of a friend, if no one else, to offer you constructive criticism about your work, and perhaps also about your personal conduct. If we did not have this form of feedback we would never improve. But there is a time and place for everything. The time is not when you are exploring and experimenting with new ideas. This is why professional creative thinkers – authors, inventors and artists, for example – rarely talk about work in progress.

Certain environments are notoriously hostile to creative work. Paradoxically, universities are among them. One of the main functions of a university is to extend the frontiers of knowledge. Therefore you would expect a university to be a community of creative scientists, engineers, philosophers, historians, economists, psychologists and so on. But academics are selected and promoted mainly on account of their intelligence, even cleverness, as analytical and critical scholars, not as creative thinkers. An over-critical atmosphere can develop.

As a leader you should watch out for the symptoms and make sure that the climate remains positive. That may not be easy if your team includes brilliant and creative, but also egotistical – and perhaps arrogant – young minds.

The role of criticism

Frances Crick joined the group studying molecular biology in the Cavendish Laboratory at Cambridge, which later formed the nucleus for the independent Laboratory for Molecular Biology. The group was under the general supervision of Sir Lawrence Bragg, a Nobel laureate for his work on X-ray crystallography.

At this time Crick was over thirty, with no research record to speak of. But he told the group that they were all wasting their time, for, according to his analyses, almost all the methods they were pursuing had no chance of success. He read them a paper – only his second research paper – entitled 'What Mad Pursuit!' – a quotation from Keats's 'Ode on a Grecian Urn'. He continues:

Bragg was furious. Here was this newcomer telling experienced X-ray crystallographers, including Bragg himself, who had founded the subject and been in the forefront of it for almost forty years, that what they were doing was most unlikely to lead to any useful result. The fact that I clearly understood the theory of the subject and indeed was apt to be unduly loquacious about it did not help. A little later I was sitting behind Bragg, just before the start of a lecture, and voicing to my neighbour my usual criticism of the subject in a rather derisive manner. Bragg turned around to speak to me over his shoulder. 'Crick,' he said, 'you're rocking the boat.'

There was some justification for his annoyance. A group of people engaged in a difficult and somewhat uncertain undertaking are not helped by persistent negative criticism from one of their number. It destroys the mood of confidence necessary to carry through such a hazardous enterprise to a successful conclusion. But equally it is useless to persist in a course of action that is bound to fail, especially if an alternative method exists. As it has turned out, I was completely correct in all my criticisms with one exception. I underestimated the usefulness of studying simple, repeating, artificial peptides (distantly related to proteins), which before long was to give some useful information, but I was quite correct in predicting that only the isomorphous replacement method could give us the detailed structure of a protein.

I was still, at this time, a beginning graduate student. By giving my colleagues a very necessary jolt I had deflected their attention in the right direction. In later years few people remembered this or appreciated my contribution except Bernal, who referred to it more than once. Of course in the long run my point of view was bound to emerge. All I did was to help create an atmosphere in which it happened a little sooner. I never wrote up my critique, though my notes for the talk survived for a few years. The main result as far as I was concerned was that Bragg came to regard me as a nuisance who didn't get on with experiments and talked too much and in too critical a manner. Fortunately he changed his mind later on.

Francis Crick, *What Mad Pursuit: A Personal View of Scientific Discovery* (Penguin, 1990)

In this passage Crick points us to one aspect of the truth about criticism. Sometimes an individual needs to be courageous in challenging accepted views, and to persist in criticism despite group pressures to conform. Such criticism may be voiced in vivid language in order that it may penetrate the thick hides of fixed ideas and win a hearing for itself. It may be consciously rejected, Crick notes, but it is yet to have influence at a subliminal level on the corporate unconscious mind of the group, perhaps even altering its direction of thought.

CHOOSE CONSTRUCTIVE CRITICS

When, as a young historian, G. M. Trevelyan told his professor that he wanted to write books on history, he was at once advised to leave Cambridge University on the grounds that the university was far too critical an environment for anyone who wished to be a writer.

The novelist Iris Murdoch gave much the same reason for leaving academic life as a philosopher at Oxford: writing creative fiction is seldom done well in the critical climate of a university.

The same principle applies to schools, colleges, churches, industrial and commercial organizations, even families. Surround yourself with people who are not going to subject your ideas to premature criticism.

'I can achieve that easily by not talking about them,' you might reply. Yes, but that cheats you out of the kinds of discussion which are generally valuable to thinkers. These fall under the general principle that 'two heads are better than one'.

- It is useful to hear another person's perspective on the problem.
- They may have relevant experience or knowledge.
- They are likely to spot and challenge your unconscious assumptions.
- They can lead you to question your preconceptions and what you believe are facts.

In short, you need other people in order to think – for thinking is a social activity – but you do not need *over-critical* people, or those who cannot reserve their critical responses in order to fit in with your needs.

'A new idea is delicate,' said Charles Brower. 'It can be killed by a sneer or a yawn; it can be stabbed to death by a quip and worried to death by a frown on the right man's brow.'

The management of criticism is almost as important as the management of innovation. Criticism has to be made. For expensive mistakes may occur, leading organizations up blind and profitless alleys, if ideas are not evaluated rigorously at the right time.

But critical evaluation should not be applied prematurely in the creative process. Sometimes ideas have to evolve quite far before any practical and commercial use becomes apparent. But tested they must be by others at various stages of their life history. The good ones are those that can then jump the hurdles of criticism.

KEY POINTS: SUSPENDING JUDGEMENT

- Suspending judgement means erecting a temporary artificial barrier between the analysing, synthesizing and imaginary faculties of your mind on the one hand, and the valuing, evaluating, criticizing and judging skills on the other hand.
- Premature criticism from others can kill seeds of creative thinking. Besides managing your own critical faculty you have to turn the critical faculties of others to good account. That entails knowing when and how to avoid criticism as well as when and how to invent it.
- Some social climates in families, working groups or organ-

izations encourage and stimulate creative thinking, while others stifle or repress it. The latter tend to value analysis and criticism above originality and innovative thinking.

Criticism often takes from the tree caterpillars and blossoms together.

Jean-Paul Sartre

CHECKLIST:
SUSPENDING JUDGEMENT

Answer Yes or No to the following:	Yes	No
Do you have a tendency to evaluate your own ideas – or half-ideas – far too soon?	☐	☐
Have you sometimes abandoned a promising idea because someone criticized it at that early stage?	☐	☐
Can you give two examples where other people have had a constructive and beneficial influence on the development of your ideas?	☐	☐

1. ....
2. ....

	Yes	No
Do you sometimes find it difficult to refrain from criticizing the ideas of other people when they first mention them?	☐	☐

List the five critical phases that, in your experience, kill ideas dead in the mind's womb (such as 'It will never work'):

1. ....
2. ....
3. ....
4. ....
5. ....

	Yes	No
Do you promise never to use these phrases again?	☐	☐

HABIT FIVE: USING THE STEPPING STONES OF ANALOGY

Metaphor, simile and analogy are integral to our language. Behind many words there are collapsed metaphors. Take for example, the words 'scruple' and 'stimulus'. In the Latin, *scrupulus* and *stimulus* meant, respectively, a small stone in your shoe and a goad or prod. All language originally stemmed from pictures and that is the way we still think. We use metaphor and simile to illustrate what we say or to adorn our meaning. But creative thinkers use metaphor and analogy on purpose: as part of this fifth habit of thinking analogically as opposed to logically. The former is concrete and employs pictures; the latter is essentially abstract and non-visual. There is a skill in analogizing.

Metaphor, which comes from the Greek verb meaning 'to transfer', usually refers to a figure of speech in which a name or descriptive word or phrase is transferred to an object or action that is different from, *but analogous to*, that to which it is literally applicable. The title of this chapter, for example, uses a metaphor.

On metaphor in daily life

'Metaphor is for most people a device of the poetic imagination and the rhetorical flourish – a matter of extraordinary language. Moreover, metaphor is typically viewed as characteristic of language alone, a matter of words rather than thought or action. For this reason, most people think they can get along perfectly well without metaphor. We have found, on the contrary, that metaphor is pervasive in everyday life, not just in language, but in thought and action. Our ordinary conceptual system, in terms of which we both think and act, is fundamentally metaphorical in nature...

'The essence of metaphor is understanding and experiencing one kind of thing in terms of another.'

G. Lakoff and M. Johnson, *Metaphor We Live By*
(University of Chicago Press, 1980)

The essence of metaphor as a tool of thought lies in that core of analogy. So what exactly is analogy? It has two overlapping meanings:

- Inference that if two or more things agree with one another in some respects they will probably agree in others
- Resemblance in some particulars between things otherwise unalike

Notice that analogy usually implies a likeness or parallelism in *relations* or *attributes* rather than in appearance. Analogizing in the first sense is a form of presumptive reasoning. For when you draw an analogy, you are acting on the reasonable, or even possible presumption (or assumption), that if things have some similar attributes they will have other similar attributes. Remember, we are doing it all the time.

HABIT FIVE: USING THE STEPPING STONES OF ANALOGY

Two mothers are waiting outside a school to collect their children at the end of the day and chatting. 'Your Jane's new boyfriend Mark is just like my brother Jack used to be, the life and soul of the party,' said one. 'They've got the same sense of humour, too. Jack was a devil with the girls, too,' she continued, 'so I don't expect it will last long.'

'Didn't you once tell me that Jack had a bit of a temper?' asked her companion.

'He certainly did. I have warned Jane to watch out for that, as I am sure it's there. Though, to tell you the truth, he has been as mild as a lamb – so far.'

As a form of reasoning, analogy has to be handled carefully. *For all analogies break down at some point.* You need to know when to jump off the train. In the story above, for example, Mark certainly is analogous to Jack in two respects. But there are no grounds for believing that the analogy will hold in the other respects mentioned.

Creative thinkers, however, are analogizing in a rather different way. For them analogies are often the source of new ideas or new ways of approaching a problem.

THE ANALOGICAL PATH TO INNOVATION

Put yourself into the shoes of an inventor. You have become dissatisfied with the solution to some existing problem or daily necessity. You are casting about in your mind for a new idea. Something occurs to you, possibly suggested by reading about other people's attempts in the files of the patent office. You go home and sketch your invention, and then make a model of it.

There are other later stages, of course, but let us stop here. The point is that the model you have reached may well have been suggested by an analogy from Nature. Indeed, you could look upon Nature as a storehouse of models waiting to be used by inventors. In a companion book to this one, *Effective Decision Making*, I included the following quiz, which you might like to attempt to answer now:

QUIZ
List specific inventions that were (or might have been) suggested to creative thinkers by the following natural phenomena:
(1) human arms   (6) earthworms   (2) cats   (7) a flower   (3) seagulls   (8) the eye of a fly   (4) a frozen salmon   (9) conical shells   (5) spiders   (10) animal bone structures

Answers on pages 209–10.

Can you add to that list? Take a piece of paper and see if you can add at least five other inventions that have sprung into the inventor's mind by using an analogy as a stepping stone.
In case you get stuck, here are some more natural phenomena that *could* have suggested inventions to alert creative thinkers. Can you identify what these inventions might have been?
(11) dewdrops on leaves   (14) human foot   (12) human skulls   (15) human lungs   (13) bamboo   (16) larynx

Answers on pages 209–10.

PUTTING AN ANALOGY TO WORK

The British industrialist Lord Weinstock, in his early days as a television set manufacturer, made one of his most significant advances through reducing the cost of making a television set case by more than half.

This particular innovation – for innovation it was – arose because a bright production engineer, John Banner, knew a Scottish maker of church pews. The church-pew man had found a way of creating certain veneers so that they could be cut up cleanly in curved shapes.

Banner saw that by using these methods, television set cases could be made on a special machine that sliced them up at the end of the line like Swiss rolls.

This illustrates two aspects of creativity: first, the use of analogy – the insight that it may be possible to chop up sets like Swiss rolls: secondly, the transfer of technology between two industries – in this case the industries were remote from each other (and the church-pew maker was not a contributor to the scientific journals), so that the transfer would never have happened had it not been for the accident of personal acquaintance.

Remember that what the natural model suggests is usually a principle that Nature has evolved or employed to solve a particular problem or necessity in a given situation. That principle can be extracted like venom from a snake and applied to solve a human problem. Radar, for example, came from studying the uses of reflected soundwaves from bats. The way a clam shell opens suggested the design for aircraft cargo doors. The built-in seam weakness of the peapod suggested a way of opening cigarette packages, a method now widely used in the packaging industry.

The case of the reluctant missile

A good example of using an analogy to solve a problem is the case of a defence industry manufacturer who produced a missile that had to fit so well into the sleeve within its silo that it couldn't be pushed in. Using the analogy of a horse that refuses to be backed into a narrow stall, but has to be led in, the solution the defence contractor eventually reached, was to *pull it in with a cable.*

The fundamental principle – that models for the solution of our problems probably already exist, we do not have to create them from nothing – can be applied to all creative thinking, not just to inventing new products. Take human organization for example. Most of the principles involved can be found in Nature: hierarchy (baboons), division of labour (ants, bees), networks (spiders' webs), and so on. If you are trying to create a new organization you will find plenty of ready-made models in human society, past or present. Remember, however, that these are only analogies. If you copy directly you are heading for trouble. More of that later.

MAKE THE STRANGE FAMILIAR AND THE FAMILIAR STRANGE

When the tribal inhabitants of the jungles of New Guinea saw an aircraft for the first time they called it 'the big bird'. Birds were familiar to them. Their first step towards understanding something totally strange or unfamiliar to them was to assume it was an unusual example of something they already knew. We assimilate the strange or unfamiliar by comparing it consciously or unconsciously to what is familiar to us.

With further experience the New Guinea tribesmen doubtless discovered that in some respects aircraft are like birds and in some respects they are not. In other words, following the 'big bird' hypothesis, noting the point where it begins to break down is a useful way of exploring and beginning to understand a new phenomenon. Use analogy, therefore, to explore and understand what seems to be strange.

Imagine for a moment that an unknown animal has been discovered deep in the jungles of South America. It is destined to replace the dog and the cat in popularity as a domestic pet within the next decade. What does it look like? What are its winning characteristics? Take some paper now and draw it, making some notes about your sketch.

Your new animal may have short silky fur like a mole. Its face may be borrowed from a koala bear and its round cuddly body from a wombat. It is blue in colour and green in temperament, for it does not foul the sidewalks or parks. That sounds a bit like a cat. It repels unwanted intruders more effectively than a guard-dog, but is as gentle with children as a white rabbit.

What you are tending to do, consciously or subconsciously, is to borrow characteristics from the animals you know. There is nothing wrong with that. For as we know, we humans cannot make anything out of nothing.

The chemistry of leadership

At a seminar for university heads of departments that touched on the differences between leadership and management, one of the participants, a professor of chemistry, used the familiar to understand the unfamiliar in this way:

> In chemistry a reaction between two compounds that can react is often put down in notation as follows:
>
> $$A + B \leftrightarrows AB$$
>
> Many reactions proceed slowly, if at all, without a *catalyst*. This to my mind is the role of leadership in getting a job done – to catalyse the process.
>
> There are various ways in which the analogy could be amplified but if you consider a rough equation of:
>
> $$\text{PROBLEM} \leftrightarrows \text{SOLUTION}$$
>
> Management will realize a solution in many instances but leadership will usually catalyze it. There is a little magic involved!

Creative thinking often involves a leap in the dark. You are looking for something new. By definition, if it is really novel neither you nor anyone else will have had that idea. You can't get there in one jump. But if you can hit upon an analogy of what the unknown idea may be like, you are halfway there.

The reverse process – making the familiar strange – is equally useful to the creative thinker. Familiarity breeds conformity. Because things, ideas or people are familiar, we stop thinking about them. As the Roman philosopher Seneca said, 'Familiarity reduces the greatness of things.' Seeing them as strange, odd, problematic, unsatisfactory or only half-known restarts the engines of our minds. Remember the saying that God hides things from us by putting them near to us.

As an exercise in warming up your latent powers of

creative thinking you can do no better than to apply this principle of making the familiar strange. Take something that you frequently see or experience, or perhaps an everyday occurrence like the sun rising or the rain falling. Set aside half an hour with some paper and a pen or pencil. Reflect or meditate on the object, concentrating on what you do not know about it.

A family member or a friend makes a good subject for this exercise. When we say we know someone we usually mean that we have a hazy notion of their likes and dislikes, together with a rough idea of their personality or temperament. We believe we can predict more or less accurately how the person will react. We think we know when our relative or friend is deviating from their normal behaviour. But take yourself as an analogy. Does anyone know everything about you? Could you in all honesty say that you fully know yourself?

'We do not know people – their concerns, their loves and hates, their thoughts,' the novelist Iris Murdoch once said in a television interview. 'For me the people I see around me every day are more extraordinary than any characters in my books.' The implication is that below the surface of familiarity there is a wonderful unknown world to be explored.

EXERCISE 2: Using the stepping stones of analogy
Think of one situation – personal or work-related – in which you would like to try using analogies to solve a problem. Then go through the following stages.

Step One: State problem

....

....

Step Two: Think of analogies (your situation is like . . .)

1. ....
2. ....
3. ....
4. ....
5. ....

Step Three: List the alternatives suggested by the analogies

....

....

Step Four: Transfer suggested solutions to problem

....

....

KEY POINTS: USING THE STEPPING STONES OF ANALOGY

- Thinking by analogy, or analogizing, plays a key part in imaginative thinking. This is especially so when it comes to creative thinking.
- Nature is a quarry for models that suggest principles for the solution of problems.
- There are other models or analogies to be found in existing products and organizations. Why reinvent the principle of the wheel when it has already been discovered? Some simple research may save you the bother of thinking it out for yourself.
- The process of understanding anything or anyone unfamiliar, foreign, unnatural, unaccountable – what is not already known, heard or seen – is best begun by relating

it by analogy to what we know already. But it should not end there.
- The reverse process of making the familiar strange is equally important for creative thinking. We do not think about what we know. Here artists can help us to become aware of the new within the old.
- 'No man really knows about other human beings,' wrote the novelist John Steinbeck. 'The best he can do is to suppose that they are like himself.'

Don't be afraid of taking a big step – you cannot cross a chasm in two steps.
David Lloyd-George

HABIT SIX: TOLERATING AMBIGUITY

If you want people to stick to the status quo, offer them plenty of alternatives. When people are faced with lots of choice, doing nothing becomes, paradoxically, more attractive. Estate agents, for example, know that if they want to make a sale, they shouldn't show you too many different houses.

> A medical researcher in Canada recently asked a group of doctors to make choices. Half of them were given two options in the treatment of a patient, while the other half were given a third alternative as well. The decision was whether or not to prescribe drug treatment. Given a straight choice between drug or no drug, 72 per cent opted for the drug. But when given a more complex choice, with two drugs to choose from, only just over half prescribed either.

In *Effective Decision Making*, I offered a sieve-like model of decision making to explain this phenomenon. We usually begin with lots of feasible possibilities of one kind or another. These are steadily reduced until they funnel down into an either/or choice.

HABIT SIX: TOLERATING AMBIGUITY

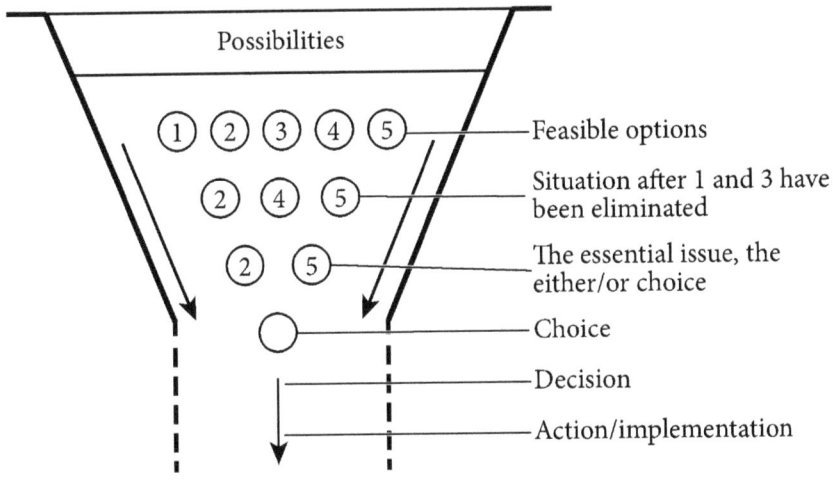

The Decision Making Funnel

As an effective decision maker, your aim – in working on the feasible options – is to reduce them eventually to two alternatives – *either* this *or* that – as soon as possible. But remember the proverb: 'More haste, less speed.'

Alternatives, in the strict sense, are mutually exclusive. The first thing to do is check if the alternatives are truly alternative. For it may be a case of both-and rather than either-or. There are situations when you can opt for both courses, possibly in sequence – the trial-and-error method. Or you can combine, mix or blend them in some other way, possibly using your powers of creative thinking.

The Funnel model poses plenty of problems for unwary thinkers. Perhaps the most common is moving too quickly to the either/or issue, so that some *feasible* options – given one's resources – are ignored or prematurely omitted. Why does it happen? Mainly because indecision and uncertainty are states that on the whole people don't enjoy. A decision – any decision – is seen to be preferable to enduring any longer the state of indecision. As the Arab proverb says, 'Men sleep well as guests in the Hotel of Decision'.

Of course, there are times when it is almost true that any decision is better than no decision. Setting aside such situations, however, the enemy of creative thinking in problem solving or decision making is a low *tolerance for ambiguity*.

Ambiguous comes from a Latin verb meaning 'to wander around'. When you are wandering around in a mist of doubt and uncertainty arising from obscurity and indistinctiveness, you are afflicted by ambiguity. Nothing is clearly defined. It's all indeterminate. You are hesitating. The ground is constantly shifting. No meanings are unequivocal. In summary, there is a *lack of clarity* that is difficult for most people to sustain for long.

EXERCISE 3: Decision making
Can you think of an example in your own life when you have made a decision in order to end a state of uncertainty?

Habit Six for the creative thinker is to be much more tolerant of ambiguity and apparent failure than their less creative colleagues. People who can do this are better at overcoming or successfully controlling the impulse to come up with solutions or make decisions *prematurely*. For the good idea can be the enemy of the best idea.

You can see why action-orientated people, who are 'doers', and those who are more creative can be at loggerheads! But so can husbands and wives. There may be a tension between the two orientations, but there is not necessarily conflict. They are potentially complementary, and certain major fields – such as industry and commerce – require both.

Develop the habit of deferring the *either/or* decision for as long as possible. Being decisive doesn't mean making staccato decisions. It means taking decisions at the right time. Then you will be able to explore the 'outer space' of the possibilities – the feasible ones and even those that may be way out or 'off the wall' but somehow interest you.

NEGATIVE CAPABILITY

'Negative Capability, that is when a man is capable of being in uncertainties, mysteries, doubts, without any irritable reaching after fact and reason.' These words by the poet John Keats point to an important attribute. Negative capability was, he felt, the supreme gift of William Shakespeare as a creative thinker. It is important, he adds, for all creative thinkers to be able 'to remain content with half-knowledge'.

Some people, as we have seen, find any sort of ambiguity uncomfortable and even stressful. They jump to certainties – any certainties – just to escape from the unpleasant state of not knowing. They are like the young man who will not wait to meet the right girl, however long the waiting, but marries simply in order to escape from the state of being unmarried.

Thinking sometimes leads you up to a locked door. You are denied entry, however hard you knock. There seems to be some insurmountable barrier, a refusal to give you what you are seeking. Yet you sense something is there. You feel as if you are in a state of suspended animation; you are wandering around in the dark. All you have are unanswered or half-answered questions, doubts, uncertainties and contradictions. The temptation to anxiety or fear is overwhelming. Anxiety is diffused fear, for the object of it is not known clearly or visibly. If you are in a jungle and see a tiger coming towards you, you are afraid; if there is no tiger and you still feel afraid, you are suffering from anxiety.

Following that analogy, what the person needs is courage. Courage does not mean the absence of anxiety or fear – it would be inhuman not to experience these emotions – but, rather, the ability to contain, control or manage anxiety, so that it does not freeze us into inaction.

Thinkers who are more creative have a higher threshold of tolerance to uncertainty, complexity and apparent disorder than others. For these are conditions that often produce the best results. They do not feel a need to reach out and pluck a premature conclusion or unripe solution. That abstinence requires an intellectual form of courage. For you have to be able to put up with doubt, obscurity and ambiguity for a long time, and these are negative states within the kingdom of the positive. The negative and the positive are always at each other's throats, so you are condemned to an inner tension.

The great American pioneer Daniel Boone, famous for his journeys into the trackless forests of the western frontier in the region we now call Kentucky, was once asked if he was ever lost. 'I can't say I was ever lost,' he replied slowly, after some reflection, 'but I was once sure bewildered for three days.' As a creative thinker you may never feel quite lost, but you will certainly be bewildered for long stretches of time. When your mind does not know where it is going, it *has* to wander around for a while.

Courage and perseverance are cousins. 'I think and think, for months, for years,' said Einstein. 'Ninety-nine times the conclusion is false. The hundredth time I am right.' Creative thinking often – not always – does require an untiring patience. Secrets are not yielded easily. You have to be willing, if necessary, to persist in your particular enterprise of thought, despite counter-influences, opposition or discouragement.

The habit of tolerating ambiguity helps to develop the qualities of courage, perseverance and patience. Those characteristics in turn will sustain you in the presence of unavoidable uncertainty. 'The bird carries the wings, but the wings carry the bird.'

SOME USEFUL STRATEGIES

The longer you are in the presence of a difficulty the less likely you are to solve it. Although creative thinking requires sustained attention, sometimes over a period of years, it does not always have to be conscious attention. It is as if you are delegating the question, problem or opportunity to another department of your mind. Having briefed your depth mind, as it were, by conscious mental work, you should then switch off your attention. Wait for your unconscious mind to telephone you. 'Hey, have you thought of this . . .?'

You should learn to expect your depth mind to earn its living. Remember, the testimonies to its capacity for creative work are overwhelming. 'My stories and the people in them', said the writer H. E. Bates, 'are almost wholly bred in imagination, that part of the brain of which we really know so little, their genesis over and over again inspired by little things, a face at a window, a chance remark, the disturbing quality of a pair of eyes, the sound of wind on a seashore. From such apparent trivialities, from the merest grain of fertile seed, do books mysteriously grow.'

A friendly and positive expectancy is rewarded when your depth mind stirs. The important thing then is to keep your analytical and critical powers switched off. 'When your demon is in charge,' said Rudyard Kipling, 'do not try to think consciously. Drift, wait and obey.'

> The British composer George Benjamin has given us a vivid picture of the creative process: 'I hate it when people describe my composing as a "gift". All people have gifts, even if they don't all realize them,' he says. 'I'm lucky enough to have been encouraged to believe in my abilities. When I'm composing I start slowly. For weeks I don't

really do anything, just walk round in circles, thinking. But that *is* the composition: the mind subconsciously sorts things out, and later on it comes pouring out – as though the piece were writing itself. An orchestral work can contain several hundred thousand notes, all relating to one another. At the beginning one is trying to determine the laws that will govern those relationships, which is intellectual rather than creative. But none of the hard work is wasted. *The mind connects things in unbelievable ways.* And at the end, it all pours out.'

The mind does indeed connect things in unbelievable ways. For Leonardo da Vinci the worlds of science and art were deeply interconnected. His scientific notebooks were filled with pictures, colours and images; his sketchbook for paintings abounded with geometry, anatomy and perspective. He wrote:

> *To develop a complete mind:*
> *Study the science of art;*
> *Study the art of science.*
> *Learn how to see.*
> *Realize that everything*
> *connects to everything else.*

Remember those words of Rodin – 'I invented nothing; I rediscover.' It may help you to have confidence if you know there *are* connections: then it becomes a matter of discerning, selecting and combining.

You may become aware that your depth mind has done some work for you when your body is active but your mind is in neutral. Ideas often come to people when they are walking or driving a car. In my earlier books I have described how the key connections that led both to the development of X-ray crystallography and to the invention of the body scanner occurred to their originators while out walking. Physical relaxation – sitting on a train, having a bath, lying awake in the morning – is another conducive state.

The novelist John le Carré is one of the many creative thinkers who find that walking plays a part in the total economy of creative thinking, albeit not a direct one. 'I have a walking appetite just as I have other appetites, and am quite frustrated if it can't be answered on demand. Moving gets me unclogged in my head,' he says. 'I almost never make a note when I'm walking and usually forget the great lines I have composed, which is probably just as well. But I come home knowing that life is possible and even, sometimes, beautiful.'

GET WORKING – DON'T WAIT FOR INSPIRATION

'I can call spirits from the vasty deep,' boasts Owen Glendower in Shakespeare's *Henry IV*. Hotspur puts down the fiery Welshman by replying, 'Why so can I, or so can any man; but will they come when you do call for them?'

Doubtless Shakespeare is writing here from personal experience. The comings and goings of inspiration are unpredictable. Magic is the human attempt to control or manage God; it never works. We cannot even earn the enriching presence of inspiration by our good works and good behaviour, like prisoners hoping for remission. We seem to be powerless.

In creative work it is unwise to wait for the right mood. 'Writing has to develop its own routine,' said Graham Greene in an interview not long before his death. 'When I'm seriously at work on a book, I set to work first thing in the morning, about seven or eight o'clock, before my bath or shave, before I've looked at my post or done anything else.

If one had to wait for what people call "inspiration", one would never write a word.'

The thriller writer Leslie Thomas would agree. 'People are always asking me, "Do you wait for inspiration?" But any novelist who does that is going to starve. I sit down, usually without an idea in my head, and stare at the proverbial blank paper; once I get going, it just *goes*.'

It can seem impossible, like trying to drive a car with more water in the tank than petrol. But you just have to get out and push. Better to advance by inches than not to advance at all.

Thomas Edison, inventor of the electric light bulb among many other things, once gave a celebrated definition of genius as 'One per cent inspiration and ninety-nine per cent perspiration.' Creative thinking, paradoxically, is for ninety-nine hours out of every hundred not very creative: it is endlessly varied combinations of analysing, synthesizing, imagining and valuing. The raw materials are sifted, judged, adapted, altered and glued together in different ways. When Queen Victoria congratulated the world-renowned pianist Jan Paderewski on being a genius he replied, 'That may be, ma'am, but before I was a genius I was a drudge.'

Not all intellectual drudges, however, are geniuses. Something more is needed that lies beyond the willingness to start work without tarrying for inspiration and to keep at it day in and day out. You also need a peculiar kind of sensitivity, as if you were standing still and waiting, prepared and ready with all your senses alert, for the faintest marching of the wind in the treetops. Your spiritual eye may trace some delicate motion in your deeper mind, some thought which stirs like a leaf in the unseen air. It is not the stillness, nor the breath making the embers glow, nor the half-thought that only stirred, but these three mysteries in one that together constitute the experience of inspiration.

The German poet Goethe uses a more homely image. 'The worst is that the very hardest thinking will not bring thoughts,' he writes. 'They must come like good children of God and cry, "Here we are." But neither do they come unsought. You expend effort and energy thinking hard. Then, after you have given up, they come sauntering in with their hands in their pockets. If the effort had not been made to open the door, however, who knows if they would have come?'

One incident in the life of James Watt illustrates Goethe's principle beautifully. Watt found that the condenser for the Newcomen steam engine, which he studied at the University of Glasgow, was very inefficient. Power for each stroke was developed by first filling the cylinder with steam and then cooling it with a jet of water. This cooling action condensed the steam and formed a vacuum behind the piston, which the pressure of the atmosphere then forced to move. Watt calculated that this process of alternately heating and cooling of the cylinder wasted three-quarters of the heat supplied to the engine. He realized, therefore, that if he could prevent this loss, he could reduce the engine's fuel consumption by more than 50 per cent. He worked for two years on the problem with no solution in sight. Then, one fine Sunday afternoon, he was out walking:

> I had entered the green and had passed the old washing house. I was thinking of the engine at the time. I had gone as far as the herd's house when the idea came into my mind that as steam was an elastic body it would rush into a vacuum, and if a connection were made between the cylinder and an exhausting vessel it would rush into it and might then be condensed without cooling the cylinder . . . I had not walked further than the Golf house when the whole thing was arranged in my mind.

'Like a long-legged fly upon the stream, her mind moves upon silence.' These evocative words by Robert Frost underline the need for silence and solitude in creative thinking, such as you find on a country walk. It helps, too, if you have a feeling of expectancy or confidence. We have all been given minds capable of creative thinking and there is no going back on that. So we are more than halfway there. We just have to believe that there are words and music in the air, so to speak, if we tune our instruments to the right wavelengths. They will come in their own time and place. Our task is to be ready for them. For inspiration, like chance, favours the prepared mind. By contrast, negative feelings of fear, anxiety or worry – even anxiety that inspiration will never come or never return – are antithetical to this basic attitude of trust. They drive away what they long for. As expressed by the poet Shelley: 'If winter comes, can spring be far behind?'

KEY POINTS: TOLERATING AMBIGUITY

- Knowing when to turn away from a problem and leave it for a while is an essential skill in the art of creative thinking. It is easier for you to do that if you are confident that your unconscious mind is taking over the baton.
- Even when ideas – or hints of ideas – are beginning to surface, resist the temptation to start thinking consciously about them. Let them saunter in at their own time and place. A heightened awareness and detached interest on your part will create the right climate.
- All creative thinking stems from seeing or making connections. Perhaps everything is connected with everything else, but our minds cannot always perceive the links. From the myriads of possible combinations, moreover, we have to select according to different criteria to our field. Is it

simple? Is it true? Is it beautiful? Is it useful? Is it practicable? Is it commercial?
- Negative Capability is your capacity to live with doubt and uncertainty over a sustained period of time. 'One doesn't discover new lands', said French novelist André Gide, 'without consenting to lose sight of the shore for a very long time.'
- It is part of a wider tolerance of ambiguity that we all need to develop as persons. For life ultimately is not clearly understandable. It is riven with mystery. The area of the inexplicable increases as we grow older.
- 'A man without patience is a lamp without oil,' said the classical guitarist, Andrés Segovia. Creative thinking is a form of active, energetic patience. Wait for order to emerge out of chaos. It needs a midwife when its time has come.

> *One should never impose one's views on a problem; one should rather study it, and in time a solution will reveal itself.*
>
> Albert Einstein

CHECKLIST:
TOLERATING AMBIGUITY

Answer Yes or No to the following:	Yes	No
Do you tend to reduce all decisions into two either/or alternatives too quickly?	☐	☐
'People who can't make decisions quickly annoy me.' Do you agree with this statement?	☐	☐

Can you list three possible side effects of living in a state of uncertainty over a problem or decision?

1. ....

2. ....

3. ....

	Yes	No
'Sometimes a good idea is the enemy of the best idea.' Do you agree?	☐	☐
Are you able to stop worrying about intractable problems and turn them over to your depth mind – with a deadline?	☐	☐
Are you consciously trying to develop your patience and perseverance as a thinker?	☐	☐
While in the state of half-knowledge, uncertainty or doubt, are you constantly on the look-out for possible connections 'beyond the Nine Dots'?	☐	☐

HABIT SEVEN: IDEAS BANKING

Imagine for a moment that your mind is like a bank. You are constantly making withdrawals in person or writing cheques. But your account will soon be overdrawn unless you are paying in money and building up your deposits. The money or cheques you pay in may seem to have no relation to the money you withdraw, but somewhere in the vaults of the bank it is put into simmering vats and boiled down into a usable currency of the mind: a collection of ideas, data or impressions that is available to you as a creative thinker.

This chapter concerns the depositing aspects of your intellectual banking. It is largely a natural process; providing your eyes and ears are open you can't help taking in information like a whale swallowing plankton. But art improves on Nature. Six more specific habits encapsulated within the habit of ideas banking are:

- Curiosity
- Observation
- Listening
- Reading
- Travelling
- Recording

This chapter is concerned with your strategic or long-term future as a thinker, rather than tips or hints on how to solve a particular problem. If you don't invest in yourself, or rather your mind, as recommended, your bank balance will slowly run down. What farmer can expect harvest after harvest if he never puts some goodness back into the soil? Take a strategic or long-term approach to stocking or equipping your mind to do its work for you, especially if you have now identified creative management as one of your priorities.

CURIOSITY

Akio Morita, co-founder of the Sony corporation, once said in an interview, 'My chief job is to constantly stir or rekindle the curiosity of people that gets driven out by bureaucracy and formal schooling systems.' Being constantly curious is what makes a good manager.

If you or I had been in Napoleon's shoes after his shattering defeat at Waterloo we might well have lapsed into a state of inward-looking depression if not despair. Not so Napoleon. Following his defeat he abdicated with the apparent intention of going into exile in America. At Rochefort, however, he found the harbour blockaded and decided to surrender himself to the Royal Navy. He was escorted aboard HMS *Bellerophon*. It was a new experience for him to see the inside of a ship of the Royal Navy, the instrument of France's defeat at Trafalgar a few years earlier. An English eyewitness on board noticed, 'He is extremely curious, and never passes anything remarkable in the ship without immediately demanding its use, and inquiring minutely into the manner thereof.'

> *Curiosity is one of the permanent and certain characteristics of a vigorous intellect.*
>
> Samuel Johnson

Such curiosity is – or should be – the appetite of the intellect. Creative thinkers have it, because they need to be taking in information from many different sources. The novelist William Trevor, for example, sees his role as an observer of human nature. 'You've got to like human beings, and be very curious,' he says, otherwise he doesn't think it is possible to write fiction.

Of course, curiosity in this sense must be distinguished from the sort of curiosity that proverbially kills the cat. The latter implies prying into other people's minds in an objectionable or intrusive way, or meddling in their personal affairs. True curiosity is simply the eager desire to learn and know. Such disinterested intellectual curiosity can become habitual. Leonardo da Vinci's motto was 'I question'.

One of the prime aims of education, it could be argued, is to develop such an inquisitive mind. 'The whole art of teaching', wrote the French novelist Anatole France, 'is only the art of awakening the natural curiosity of young minds for the purpose of satisfying it afterwards.'

'Curioser and curioser!' cried Alice in Wonderland. Too often it is *only* something curious, rare or strange that arouses our curiosity. But what excites attention merely because it is strange or odd is often not worth any further investigation. We do have to be selective in our curiosity.

'Thinking is trying to make up the gap in one's education,' writes the philosopher Gilbert Ryle in *On Thinking* (Blackwell, 1979). It is not, of course, a matter of teaching yourself something that you want to know; you cannot teach it because you do not know it. 'What I am trying to think out for myself is indeed something that the Angel Gabriel conceivably might have known and taught me instead,' continues Ryle, 'but it is something that no one in fact did teach me. That is why I have to think. I swim because I am not a passenger on someone else's ferryboat. I think, as I swim, for myself. No one else could do this for me.'

EXERCISE 4: How curious are you?

What do *you* feel most curious about at present? List your top three subjects or people in the following categories:

	1.	2.	3.
Historical person			
Country			
Trade or profession			
Person known to you			
Hobby or activity			
Words			
Other			

SHARPEN YOUR OBSERVATION

'I am fascinated by the principle of growth: how people and things evolved,' said the portrait painter Graham Sutherland in an interview. He added that he always tried to pin down the essence of the people he painted. 'I have to be as patient and watchful as a cat.' He could see in the human face the same sort of expression of the process of growth and struggle as he found in rugged surfaces of boulders or the irregular contours of a range of hills. 'There are so many ideas I want to get off my chest. The days aren't long enough,' he added.

It may seem odd to think of painting a picture as a means of getting an idea off your chest. But for the artist the act of careful analytical observation is only part of the story. Ideas and emotions are fused into the paint in the heat of inspiration. What the artist knows and feels is married to what he sees, and the picture is the child of that union. 'Painting is a blind man's profession,' said Picasso. 'He paints not what he sees, but what he feels, what he tells himself about what he has seen.' That principle holds true not only for the kind of art for which Picasso is famous but also for the more realistic work of painters such as Graham Sutherland.

EXERCISE 5: Developing your senses

Creative people usually use their senses like antennae. Most of us tend to favour one sense (usually vision) at the expense of others. Here are some warm-up exercises for developing all your senses:

Look
- The next time you are in a garden, see if you can identify twenty different colours.
- Try writing a description of the place where you holidayed last.

Listen
- List five different sounds you can hear in the next minute.
- How many sounds can you identify in a busy shopping centre?

Touch
- Close your eyes and touch the objects around you. For example, see if you can feel the differences in different papers.

Taste	• Try distinguishing two white wines blindfolded. See if you can build up an anthology of favourite tastes.
Smell	• Imagine and recall your favourite smells. With what places are they associated?

An observation made through the eyes will undergo transformation to varying degrees in the creative mind as it is combined with other elements into a new idea, bubbling away in a cauldron of animated interest. But the observation itself needs to be clear, accurate and honest. Like a good cook, a creative thinker should work from the best materials.

Laurence Olivier was an actor renowned for his ability to build character in a creative way. 'I am like a scavenger,' he said, 'I observe closely, storing some details for as long as eighteen years in my memory.' When invited to play the title role in Shakespeare's *Richard III*, he drew upon his recollection of Jed Harris, a famous Broadway producer of the 1930s under whom he had had a bad experience. Harris had a prominent nose, which Olivier borrowed for the role, along with other features such as the shadow of the Big Bad Wolf which he had seen long ago in Walt Disney's film *Pinocchio*. Remembered films often gave him such ideas. The little dance he did while playing Shylock came from Hitler's jig for joy when France signed its capitulation in 1940, a moment shown on German newsreels.

Observation is a skill. At the lowest level it implies the ability to see what is actually in front of you. As scientists know, this is not as easy as it sounds. It is almost impossible to be totally objective. We tend to see what we know already. That does leave some creative possibilities. For, as Gustave Flaubert wrote, 'There is an unexplored side to everything, because instead of looking at things with our eyes we look at them with the memory of what others have thought.'

EXERCISES 6 and 7: Testing your observation skills

6. Select one area in your work responsibilities for special attention in the next week, such as the layout of goods in a shop or the pattern of customer calls. Simply observe and collect data on it, like a scientist studying the seashore or butterflies. Don't attempt to draw any conclusions, for the object of the exercise is solely to increase your powers of observation.
7. The next time you go to a railway station make a list of five things you have never seen before.

Our minds are programmed to notice certain things rather than others, not least by our particular interests. A botanist, for instance, will be likely to notice plants. If we see things or people repeatedly, we hardly observe them at all unless there is some change from the familiar or predictable, some deviation from the norm, which forces itself upon our attention. A good observer will be as objective as possible. Inevitably he or she will be selective in observation, guided by some idea or principle on what to look for. But, being serendipitous, you should be sensitive to what you have not been told – or told yourself – to look for.

One of the best forms of training in observation is drawing or sketching. Take some paper and a pencil and look at any object. Now *select* from what you see the key lines that give you its essential shape. You are now exercising careful analytical attention. Don't worry if you can't reproduce the object like a trained artist. That is not your aim here. You are using sketches as a means of learning to use your eyes, so that you can really see the world around you.

Such sketches, however rough and ready, will not only increase your awareness of the world but they will also help you to etch the scene in memory. In his autobiography

A Millstone Round My Neck (Methuen, 1981), the artist Norman Thelwell made this point:

> It may be that one's awareness of the world is heightened during the process of recording visual things with pencil, pen or brush. Sketchbooks and paintings, even the slightest notes, can recall not just the day and place but the hour, the moment, the sounds and smells that would have gone forever without them. I have drawings still which I did as a child and I can remember when I come across them what my brother said to me, what my mother was doing at the time, what was on the radio when I was working and how I felt about the world that day.

About 70 per cent of the information we use comes through our eyes. Develop your ability to see things, therefore, and make detailed observations. For they are the materials for future creative thinking.

LISTENING FOR IDEAS

'You hear not what I say to you,' says the Lord Chief Justice to Shakespeare's Falstaff.

'Very well, my Lord, very well,' replies the irrepressible old rogue. 'Rather, if you will excuse me, it is *the disease of not listening*, the malady of not marking, that I am troubled with'.

Poor listening ability is a common affliction, but creative thinkers do not suffer from it. Although we know very little about Falstaff's creator, we can at least be sure that he was a good listener.

What constitutes such a rare beast as a good listener? First, a good listener will have curiosity, that all-essential desire to learn. That requires a degree of humility, the key to have an open mind. For if you think you know it all, or at least if you believe you know more than the person you are talking to you are hardly inclined to listen.

Having an open mind does not guarantee that you will buy the idea, proposition or course of action being put to you. But it does mean that you are genuinely in the market place for new ideas. You will buy if the price is right. The next requirement is to control your analytical and critical urges. For your first priority is to grasp fully what the other person is actually saying. Remember the Turkish proverb: 'Listening requires more intelligence than speaking.'

READING

Reading without reflecting has been compared to eating without digesting. One page, or even one paragraph properly digested will be more fruitful than a whole volume hurriedly read. Or, as the film mogul Sam Goldwyn said to a hopeful author, 'I have read part of your book all the way through.' When you come across significant parts – the passages that speak to you – it is worth remembering the counsel of the Book of Common Prayer: 'Read, mark, learn and inwardly digest.'

> **EXERCISES 8 and 9: Testing your reading skills**
> 8 What is the most stimulating book you have read within the last two years? Do you underline key passages in books? When you read a book do you see it as a mass of printed words or a collection of ideas – some big and some little, some familiar to you and some new?

9 Go to your nearest public library with a friend and find the main non-fiction section. Ask your friend to blindfold you and to guide you round the shelves. Select any three books at random. At home, list the core ideas in each book – not more than five in each instance.

The power of a good book lies in the intimate relationship between author and reader. It is a transaction that takes place in solitude. It invites you to think for yourself about some subject away from the context of other people. The author should be able to lead you to nourishing food or refreshing water, and, though he or she cannot make you drink, he or she should provide you with plenty of encouragement to do so. These almost unique conditions of inner dialogue enable a good book to reach deep into your consciousness.

You don't have to plod through a book from page one to the end. You can skip and skim. 'I am not a speed reader,' said science fiction writer Isaac Asimov. 'I am a speed understander.' Taste the contents, then select what you want to chew and swallow. Never swallow first, for if you believe everything you read it is better not to read.

Reading books, then, can stimulate and develop your powers of creative thinking. If nothing else, a good book can put you into a working mood. If you are resolved to devote as much time and attention to your mental fitness as the average person spends on that more wasting asset, the human body, you will be inclined to explore the world of books.

TRAVELLING

Travelling is like gambling; it is ever connected with winning and losing, and generally where least expected we receive more or less than we hoped for.

Goethe

Here are two practical examples of how travel can contribute to new ideas and innovation in business. In the 1920s Simon Marks visited America and studied what was happening in the retail trade. American farmers, he found, had little time or opportunity to go shopping. Instead, they subscribed to Sears catalogues that gave them the novel guarantee of 'your money back and no questions asked'. Providing the service also meant creating new human skills; for example, by organizing suppliers to achieve new standards of efficiency. In the 1920s, Sears further adapted itself to the vast new urban market by placing its stores on the outskirts of the cities. When Marks returned from America in 1924 he followed the Sears example, and remodelled Marks and Spencer to provide high-quality goods at low prices, together with a 'money back' guarantee.

Sir Terence Conran became dissatisfied with British furniture shops some decades later. Why is buying furniture so boring, such a chore? Why do these shops look so bad? Why are the sales people there so dissatisfied? These are some of the questions he asked himself. 'The inspiration to change things,' he says, 'came from asking myself what I had seen on my travels that worked well. For me the markets of France and the ironmongers' shops in France all had the sort of qualities I found fascinating and exciting. Bringing these ideas together with the market needs in Britain led me to open the first Habitat shops.'

Use travel as a means, in other words, to feed and stimulate your curiosity about life. Show a willingness to take some risks. Develop a zest for exploration.

'The whole object of travel is not to set foot on foreign land,' wrote G. K. Chesterton. 'It is at last to set foot in one's own country as a foreign land.' The creative thinker has to turn himself into a stranger in his own home environment if he is to think what nobody has thought. Travel – more adventurous travel – is one way of doing just that.

EXERCISE 10: What sort of a traveller are you?
How adventurous are you when it comes to travel? Just look back over the last five years and list your travels, including holidays and business trips:

1. ....
2. ....
3. ....
4. ....
5. ....
6. ....
7. ....
8. ....
9. ....
10. ....

Do they suggest to you that you have a habit of mind that is interested in exploring the new or untried? Now write down three ideas for travel in the next five years which focus on new experience:

	EXPERIENCE	COUNTRY/PLACE
1.		
2.		
3.		

The secret of Japanese success

As a special bonus for reading this book, here's a remarkable secret I came across by serendipity when speaking at an international conference in Malaysia. The Japanese speakers and I became very friendly, and after one convivial evening I felt bold enough to ask them, 'Why do you allow so many Western visitors to tour your factories and steal your *kaizen* techniques for incremental product improvements? Why do you allow us to study your phenomenal innovation? We are competitors, you know.' My Japanese friend smiled and replied that the answer was very simple. At my request he wrote it down for me. As a world exclusive, here it is:

Should your Japanese not be up to it, I have supplied a translation on page 125.

RECORDING

The trouble with experience is that we forget so much of what we learn, especially if we aren't gifted with a retentive memory. In fact we all have to forget quite a lot if we are going to live and learn. To be condemned to remember everything would be a terrible torture and a real barrier to learning new things. But how do we ensure that we bank ideas or material that could be useful or valuable to us in the unknown future?

> '"The horror of the moment," the King went on, "I shall never, *never* forget!"
> '"You will, though," the Queen said, "if you don't make a memorandum of it."'

This advice from Lewis Carroll in *Alice in Wonderland* certainly applies in the field of creative thinking. The philosopher Thomas Hobbes always kept a notebook at hand. 'As soon as a thought darts,' he said, 'I write it down.' One practical step you can take now is to buy a new notebook to record possible materials for your present or future use: ideas, a scrap of conversation, something seen or heard on television or radio, a quotation from an article or book, an observation, a proverb. Write it down!

> *A person would do well to carry a pencil in their pocket and write down the thoughts of the moment. Those that come unsought are commonly the most valuable and should be secured, because they seldom return.*
>
> Francis Bacon

You have probably had the experience of waking up in the middle of the night with an idea. It was such a good one that you told yourself to remember it next morning. But, like the memory of your dreams, it fades fast away. Keep a pencil and pad by your bed. Carry a pocket notebook so that ideas that strike you while waiting for someone or travelling in the car can be recorded. You can transfer these jottings to your main notebook later.

There are two important principles in keeping a notebook as a tool for creative thinking. First, put down entries in the order in which they occur to you. Give a short title to each entry, and perhaps a date. Don't try to be too systematic, by putting everything on cards or loose-leaf paper arranged alphabetically, indexed and cross-indexed. This might be the right way to record things if you're a scientist. But it's not the right way when it comes to developing your powers as a creative thinker.

The second principle is to let your instinct or intuitive sense decide what you think is worth noting down. Include whatever impresses you as stimulating, interesting or memorable. At this stage it doesn't matter too much if the idea is right or wrong, only that it is interesting. Later – months later – you may need to do some editing, but initially what matters is whether or not the prospective entry gives you a spontaneous reaction of excitement or deep interest. As Shakespeare wrote: 'No profit grows where is no pleasure taken; in short, study what thou dost affect.'

> A commonplace book contains many notions in garrison, whence the owner may draw out an army into the field on competent warning.
>
> Thomas Fuller

Don't look at your entries too often. In my experience, the best time to browse through them creatively (unless, that is, you are hunting for a reference for a specific purpose) is on a train or air journey, waiting in airports, or on holiday when the mind is fresh and unencumbered with daily business.

THE JAPANESE SECRET TRANSLATED

'... YOU WON'T DO IT!'

It reminded me of Rudyard Kipling's words:

> They copied all they could follow,
> But they couldn't follow my mind;
> And I left them sweating and stealing,
> A year and a half behind.
>
> From *The Mary Gloster*

KEY POINTS: IDEAS BANKING

Curiosity

- Curiosity is essentially the eager desire to see, learn or know. It is the mind on tiptoe.
- Creative thinkers tend to have a habit of curiosity that leads them to give searching attention to what interests them.
- Thinking is a way of trying to find out for yourself. If you always knew where it was taking you there would be nothing to be curious about.

Observation

- The ability to give careful, analytical and honest attention to what you see is essential. If you don't notice and observe you will not think.
- Observation implies attempting to see a person, object or scene as if you had never seen it before in your life.
- The act of observation is not complete until you have recorded what you have seen, thereby helping to commit it to memory.

Listening

- A childlike curiosity and an open mind, backed up by sharp analytical skills and a sensitive judgement, are the essential prerequisites for being a good listener.
- Your priority must always be to achieve a grasp of the nature and significance of what is being said to you. Ask questions to elicit the full meaning. Understanding comes before evaluation.
- Listen out for ideas, however incomplete and ambiguous, as well as for potentially relevant facts and information.

Reading

- Nothing is worth reading that does not require an alert mind, open and eager to learn.
- Books are storehouses of ideas, thoughts, facts, opinions, descriptions, information and dreams. Some of these, removed from their setting, may connect to your present (or future) interests as a thinker.
- 'Reading is to the mind what exercise is to the body,' wrote the playwright Sir Richard Steele. Poetry and good prose – fact or fiction – require the use of your imaginative and recreative powers. Therefore they provide you with an enjoyable way of developing those faculties.

Travelling

- By travelling adventurously you will encounter the strange and see connections between it and what is already known or familiar to you, thereby enlarging your knowledge. It cures you of the illusion that the limits of your field of vision are the limits of the world.
- Travel helps to convert the familiar into the strange. Who really knows their own country if they never venture beyond its borders?

Recording

- Keeping a notebook is more than a useful habit: it is a vitally important tool for all creative thinking purposes.
- Writing down a quotation or passage, fact or piece of information, is a means of meditating upon it and appropriating it personally so that it grows into part of you.
- Imagine that your notebook is like a kaleidoscope. At a time when you are feeling in a creative frame of mind,

give it a metaphorical shake. You can play with new combinations and interconnections. They may suggest new ideas or lines of thought.

> *The use of reading is to aid us in thinking.*
> Edward Gibbon

CHECKLIST:
IDEAS BANKING

	Yes	No
Answer Yes or No to the following:		
As a result of reading and reflecting on this chapter, can you identify and list on paper three ways in which you can improve your curiosity?	☐	☐

In the next three months you will most probably sit next to a total stranger at a meal. What five questions will you now ask them?

1. ....
2. ....
3. ....
4. ....
5. ....

	Yes	No
Can you think of a manager who is more observant than you are? What beneficial results have stemmed from his/her observations?	☐	☐
Has anyone described you as a good listener within the last twelve months?	☐	☐
Are you an *active* listener, using questions like tools to prize pearls out of reluctant shells?	☐	☐
Does reading books or articles play an important part in keeping your mind stimulated and in shape?	☐	☐
Do you read fiction to develop and extend your imagination?	☐	☐
Have you ever travelled in search of ideas on how to do your job better?	☐	☐
Do you choose holidays in places that stimulate and refresh your mind as well as your body?	☐	☐

PART THREE

MANAGING FOR INNOVATION

Innovation is the lifeblood of any organization today. Nothing holds a company back – and the individuals working in it – more than a lack of interest in positive change. You cannot stand still: you either go backwards or forwards.

Innovation requires a blend of new ideas, teamwork and leadership. Apart from creativity and the ability to get things done, however, it also calls for a sound commercial or entrepreneurial sense. Innovation should be customer-driven as well as ideas-driven. The success of innovative projects, therefore, depends upon both your individual characteristics and the climate or orientation of the whole organization.

By the time you have read and completed Part Three you should have:

- A clear understanding of how you can **manage for creativity** in any organizational setting.
- A grasp of the essentials or key characteristics of organizations that build and maintain over a long

period of time an **organizational climate** fostering innovation.
- Knowledge of the main **methods** – notably brainstorming – for stimulating and harvesting new ideas in organizations, together with some suggestions for helping to bring those ideas to market.

10
HOW TO MANAGE INNOVATION

'Not geniuses but ordinary people', said the writer John Collier, 'require profound stimulation, incentive towards creative effort, and the nurture of great hopes.'

Don't blame your staff for lack of interest or lack of new ideas. There are no unmotivated or uninnovative people at work, only poor managers. Your task, as John Buchan once said, is 'not to put greatness into humanity but to draw it out, for the greatness is there already'. Transforming individual groups and organizations so that they realize the greatness within them is your great challenge as a leader. Perhaps it is also one of the main paths that leads towards your personal self-fulfilment.

Unlike the production of goods on an assembly-line, the process of innovation isn't visible or concrete. It is, nonetheless, clearly identifiable. I suggest that it falls into three main parts. These three major elements or phases are captured in the table over the page.

What is the common factor in these acts or phases? *Teamwork.* A team is a form of work group in which the members possess complementary skills that fit together like a jigsaw puzzle and create synergy – the confined action that is greater than the sum of the parts taken independently.

INNOVATION

CREATIVE THINKING MAKES IT POSSIBLE
TEAMWORK MAKES IT HAPPEN

THREE PHASES OF INNOVATION	
Generating ideas	Involving individuals and teams in producing ideas for improving existing products, processes and services – and creating new ones.
Harvesting ideas	Again involving groups of people in the gathering, sifting and evaluating of ideas.
Developing and implementing these ideas	Once more involving teams in the work of improving and developing the idea right up to the first response from a delighted customer.

LEADING INNOVATIVE TEAMS

Teams need leaders. What are your core responsibilities as a leader?

'A picture is worth a thousand words.' Your role as a leader can be summarized in the Three Circles model:

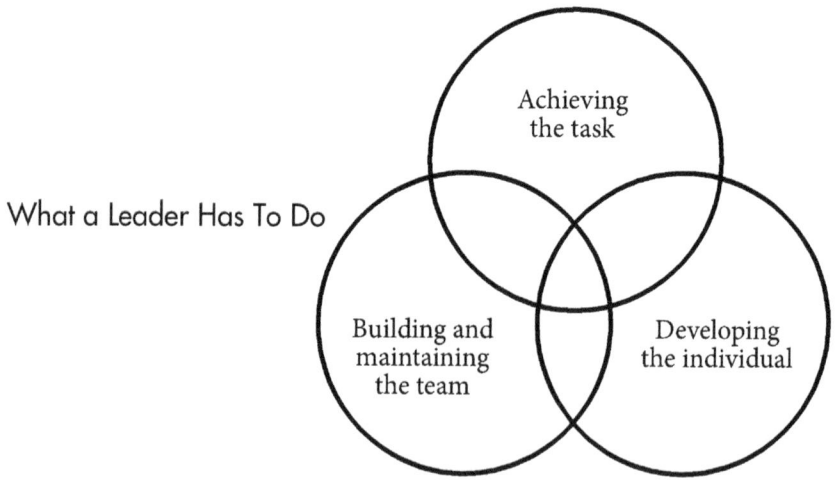

What a Leader Has To Do

HOW TO MANAGE INNOVATION

To fulfil the three circles of responsibility certain key *functions* have to be performed. They are the responsibility of the leader, but that does not mean the leader will do them all himself. They can be shared or delegated in all sorts of ways.

The following list is by no means definitive – the sheer variety of situations prohibit that – but these general functions are commonly required: It must be stressed again that

THE CORE FUNCTIONS OF MANAGERIAL LEADERSHIP	
Defining task	Correctly specifying what needs to be accomplished and breaking this task down into its discrete parts.
Planning	Formulating an effective method for achieving the task(s), i.e. organizing people, materials, time and resources in such a way that the objective(s) can be met.
Briefing	Allocating tasks and resources to subordinates in such a way that each person (a) knows what is expected of him and (b) understands the importance of his contribution.
Controlling	Keeping things to plan. Being sensitive to problems and delays and quick to respond to them. Coordinating the work of the team.
Evaluating	Making accurate and insightful judgements about proposals, past performance and people.
Motivating	Creating and maintaining the team's commitment to, and interest in, the task.
Organizing	Creating a structure and hierarchy appropriate to the task.
Setting an example	Exemplifying the values and behaviours he/she wishes to see in others.
Supporting	Encouraging group/individuals; building and maintaining good team spirit.

not all these functions will be performed by every leader all the time. In groups of more than three or four, too many actions will be needed to meet the requirements of task, team and individual for any one person to do them. But the leader is *accountable* for the three circles. Taken together, these functions constitute the generic role of leadership. That is now *your* role.

To deal effectively with people you must take time to understand them as individuals. They need to be understood both in terms of what they share in common and what differentiates them. How does this particular person differ from others? This question is clearly a vital one in the context of creative thinking and innovation.

You can begin to apply the Three Circles model in the context of innovation simply by asking the basic question, *What is the common task?*

As an experiment, try the discipline of reducing the answer to two words, a verb and a noun. A pencil, for example, 'makes marks'. This breaks habitual patterns, pulls the thinking back to fundamentals, and puts the emphasis on function rather than 'the way we've always done it'.

It may not always be easy to apply that discipline to your organization, however. If you make it too general you run the risk of losing sight of your particular niche of excellence. If you make it too specific, on the other hand, you may eliminate areas for creative development and innovation.

> An American gardening products and services company remained undecided for almost a year between two core mission statements: 'to make fertilizers' or 'to keep lawns green'. They finally chose the second. It led the company to make investments in facilities to produce a variety of chemicals and implements to keep lawns green. Such product diversification would not have been consistent

with their traditional assumption that they were in the business of producing fertilizers.

The nature of any task today has changed. It now includes – implicitly or explicitly – the demand or requirement to do it better, faster, less expensively. The need for change and innovation is thus now built in to whatever you do as your business.

> When Tom Turner took over as production manager at Stickton Lacey Ltd, a long-established company making leather goods, he found the task expressed in terms of meeting production targets. He redefined it to include '*improving* the quality and cost-effectiveness of the product'.

But creating and sustaining innovation requires extra, over and beyond the typical actions as listed above in The Core Functions of Managerial Leadership. If I'm honest, they really describe what you must do to lead and manage today's business. In order to create innovative teamwork – the key to tomorrow's business – the further set of activities required can be boiled down to five essentials:

- Electing creative people
- Encouraging creativity in teams
- Building on ideas
- Communicating about innovation
- Overcoming the obstacles that separate new ideas from the market place

SELECTING CREATIVE PEOPLE

Creative people are attracted by creative environments and opportunities. Therefore, as a manager you need to make sure that your organization has the right colours to attract them and that you advertise these attractions in the job market or to individuals you have identified.

> **EXERCISE 11: Selecting the creative candidate**
> You are about to interview someone for a job that especially requires the ability to have new ideas. It requires a fresh mind and an innovative spirit. Place the following personal qualities or attributes in order of importance:
>
> Numeracy .... A good analytical mind ....
> Lack of interest in small A flexible mind ....
>   details .... Scepticism ....
> Curiosity .... Orientation to
> Verbal skills ....   achievement ....
> Sensitivity .... Humour ....
> Wide-ranging interests ..... Persistence ....
> Enthusiasm .... Self-confidence ....
> Independence .... Non conformity ....
>
> Add *three* characteristics that you feel are missing from the list.

When the famous explorer Dr David Livingstone was working in Africa, a group of friends wrote: 'We would like to send other men to you. Have you found a good road into your area yet?'

According to a member of his family, Livingstone sent

this message in reply: 'If you have men who will only come if they know there is a good road, I don't want them. I want men who will come if there is no road at all.'

The first step in any form of team building is to choose the right people. That is a vital principle to bear in mind if you want to encourage innovation – and sustain it. Like Livingstone, in his inimitable way, you should develop an eye for the more adventurous and more independently minded person.

Any innovative organization must have a bias towards attracting intelligent and creative young people. Of course, intellectual qualities are not enough, for industry needs *doers* – people who can make things happen – rather than *thinkers* as such. There are plenty of good ideas around. The real issue is whether or not you have the people in your team or organization who are willing to put new ideas to work; in other words, to innovate. 'Give me the young man', said Robert Louis Stevenson, 'who has brains enough to make a fool of himself.'

Knowing your own strengths

Creative thinkers are clearly stronger in synthesizing, imagining and holistic thinking than others. But the best of them are equally strong in analysing ability and the faculty of valuing or judging. It is this combination of mental strengths, supported by some important personal qualities or characteristics, that make for a formidable creative mind.

General creative qualities are usually easy to identify. Creative people tend to be more open and flexible than their less creative neighbours. They bring a freshness of mind to problems. They have usually exhibited the courage to be different and to think for themselves. They are also comparatively more self-motivated and are often addicted to their work. Opposite is an expanded list of seven characteristics to look for in studying references, biographical data or during interviews.

Creative people can usually be recognized as having a pattern of characteristics represented in the list above. Such individuals don't always make natural 'company people', so your organization will need a certain psychological maturity to recruit them in the first place. Creative people can make uncomfortable companions, but can you do without them? Above all, you need teams with creative individuals who are able to lead.

Bear in mind that if you do recruit or select people with above-average creative ability for your team or organization you will find that they tend to be looking for certain compatible characteristics in you and your organization. Selection is – or ought to be – a two-way process. Before taking on creative people, check that you have the right environment for their talents to flourish. It's not much good hiring people who are only going to become frustrated. Ask yourself, What are their expectations?

Research has some clear messages on this score. It has identified the most important environmental factors in stimulating or encouraging creativity. In order of importance they are as follows:

- *Recognition and appreciation* Because the results of creative work are often postponed for a long time (many geniuses in history received no recognition in their lifetimes), creative people stand in special need of encouragement and appreciation. The recognition of the value or worth of their contribution is especially

QUALITIES OF CREATIVE PEOPLE	
General intelligence	Powers of analysing, synthesizing and valuing, as well as the ability to store and recall information.
High self-motivation	A high degree of autonomy, self-sufficiency and self-direction. Creative people enjoy the challenge. They like to pit themselves against problems or opportunities in which their own efforts can be the deciding factor. 'There is no greater joy in life,' said the inventor Sir Barnes Wallis, 'than first proving that a thing is impossible and then showing how it can be done.' Tend to be vocational in their attitude to work.
Negative capability	The ability to hold many ideas – often apparently contradictory ones – together in creative tension, without reaching for premature resolution of ambiguity. Hence, the capability of occasionally reaching a richer synthesis.
Curiosity	Sustained curiosity and powers of observation. Creative-minded people are usually good listeners.
Independence of mind	Marked independence of judgement. Resilience in the teeth of group pressures towards conformity in thinking. Seeing things as others do, but also as they do not. Thinking for oneself; thinking from first principles, not getting it out of books.
Ambiversion	Ambiverts occupy the middle ranges on the spectrum that has introversion at one end and extroversion at the other. Slightly more introvert, if anything, but need contacts with stimulating colleagues.
Wide interests	A broad range of interests, including usually those with a creative dimension.

important to them, particularly if it comes from those whose opinions they respect.
- **Freedom to work in areas of greatest interest** While the predominantly analytical person concentrates and focuses down, the creative person wanders in every possible or feasible direction. Freedom to move is the necessary condition of creative work. A creative person tends to be most effective if allowed to choose the area of work, and the problems or opportunities within that area, which arouse deep interest.
- **Contacts with stimulating colleagues** 'Two heads are better than one,' says the ancient Greek proverb. Creative people need conversation with colleagues in order to think, not merely for social intercourse.
- **Stimulating projects to work on** Along with a congenial and appreciative environment and the opportunity for appropriate recognition by their professional peers inside and outside the organization, stimulating projects or problems are especially attractive.
- **Freedom to make mistakes** Errors are inescapable in innovative work. The climate should be such that they are not all used to inflict immediate and permanent damage on one's professional career.

ENCOURAGING CREATIVITY IN TEAMS

'A camel is a racehorse designed by a committee.' That oft-quoted saying reminds us that groups have their limitations when it comes to creative thinking and decision making. But teams can think effectively together. They have a combined breadth of experience, a range of knowledge and a set of skills that is bound to be greater than that of any one individual member.

If you are setting up a project group, for example, it may very well be concerned with innovation in some form or other. The nature of the common task will guide you in the selection of the talents or abilities you need to have in the team in terms of professional or technical competencies. But in a team people also contribute as individuals. That takes us to the two sets of questions you have to ask yourself about any team member:

- Does this person share sufficiently the values, characteristics and interests of other members of the team? Will he or she work in the team harmoniously? Will the 'chemistry' work?
- What uniquely different quality does this person bring to the party? What are their particular strengths of mind? In what phase of innovation – idea generation, judgement or application – will they play a leading part?

Deliberately structuring groups in this way does pay off. It's not much good having a football team composed entirely of natural defenders. A high-performance team will consist of high-performance individuals using their different natural and developed strengths synergistically in pursuit of the common aim.

However well you select your team (providing you have discretion or a say in that key activity), it still comes back to your leadership. Does your team score goals? Or, in the case of innovative work, does it have new ideas that get implemented?

As a manager, you can do a great deal to develop the necessary team synergy by:

- Identifying the different mental skills within the team – analysing and logical, synthesizing and analogical, imaginative and holistic, judgement and wisdom.

- Making it legitimate to have a conflict of ideas or minds as long as all remain responsible for maintaining a harmony of relationships.
- Giving recognition for *any* significantly helpful contribution done well – 'It was Sally's challenge to our fixed assumption that all sales had to be paid for in sterling or dollars that led to our breakthrough.' By this means you will introduce team members to the notion that we all have different talents in different strengths but – like musicians in an orchestra – each of us can contribute to the production of the symphony.
- If necessary, teaching the team to use one or more of the idea-generating techniques, such as brainstorming (see next chapter). Apart from achieving results, these techniques help to introduce the differences of divergent thinking ('think up twenty new uses for a hammer'), to convergent thinkers, who are better at focusing down or converging on problems. It works the other way round too – it's the convergers or vertical thinkers that do best in the preliminary work of defining problems and the shaping necessary to make new ideas practicable.
- Creating an atmosphere of openness where people can speak what is in their minds – succinctly and to the point, but without fear of your displeasure on that score. Communicate quite openly about the problems and opportunities facing the team, about successes and failures. If you don't show them your cards they can't help you play your hand.

The essence of the matter lies in attitudes. It is a question of moving the team away from the position where the natural response to a half-baked idea is a negative or critical one. The expression of a more positive attitude and the surest sign of team synergy is the observed willingness to build on ideas, as shown opposite.

HOW TO MANAGE INNOVATION

BUILDING ON IDEAS

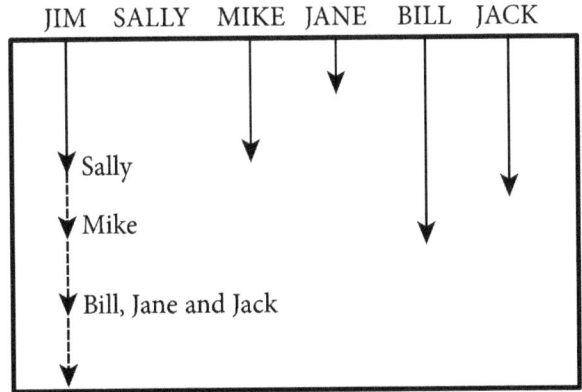

Building on Ideas

Let's visualize two meetings. In the first, a number of suggestions are made or ideas put forward for consideration, represented by the hard downward lines in the diagram. These disappear into the pond with a plop. Some of these plops might, of course, have the seeds within them of new ideas. The meeting breaks up with nothing achieved.

At a second meeting, however, after some discussion of the five original suggestions, Sally builds on or develops Mike's suggestion, thus taking it further – like a rugby ball towards the try-line. Mike then builds on *their* suggestions. Finally it's mainly Bill, Jane and Jack who put the finishing touches to what has now become the group's winning idea. Jim's half-idea has proved to be the seed, in this case, of a new product or service.

The model then is not unlike an aerial photograph of a game of rugby football, with tries being scored at the bottom end. The group at its first meeting scored no tries. By working as a team, however, in their new colours, they became remarkably successful.

The secret lies in changing attitudes and moving from the

negative or critical mode of thinking into the positive or constructive one. As Winston Churchill said at one cabinet meeting, in some exasperation at his timorous colleagues, 'Every fool can see what is wrong. See what is good in it!'

CHECKLIST: BUILDING ON IDEAS

	Yes	No
Do the group members establish a common understanding of the problem, based upon a careful diagnosis?	☐	☐
Do they focus together on a single aspect of the problem?	☐	☐
Do they actually work as a team, building on each others' ideas, or as a group of individuals?	☐	☐
Do the members take pains to make sure everyone understands each idea?	☐	☐
Is the technical content of the discussion at a high level?	☐	☐
Does anyone use analogies to suggest possible solutions?	☐	☐
Do the members really listen to one another?	☐	☐
Do the members tend to shoot down ideas quickly?	☐	☐
Does the group insist that each idea be a complete solution?	☐	☐
Or do they support and improve on an unsatisfactory idea?	☐	☐
Do they thoroughly explore one idea before going on to the next?	☐	☐
Do people keep to the point and not waste time?	☐	☐

The key principle of team synergy – building on other people's ideas and contributions – is not confined to meetings. As mentioned earlier, new ideas are sometimes born prematurely. It is often because the technology to realize them or to make them commercial is lacking that they fail. Equally often the seeds of ideas may fall on parched and dry soil. There they must await the warmth and rain of someone else's attention before they can take root.

Idea transplants

Many ideas grow better when transplanted into another mind than in the one where they sprang up.
Oliver Wendell Holmes Jr

Sometimes this more extended process of building on ideas takes place within a single innovative organization over a period of time.

> The invention of Scotch Tape in 1930 is a highlight in the story of 3M, the Minnesota corporation that grew from being a maker of mediocre sandpaper into an international conglomerate. A salesman who visited the auto plants noticed that workers painting new two-toned cars were having trouble keeping the colours from running together. Richard G. Drew, a young 3M lab technician, came up with the answer: masking tape, the company's first tape.
>
> In 1930, six years after DuPont introduced cellophane, Drew figured out how to put adhesive on it, and Scotch Tape was born, initially for industrial packaging. It didn't really begin to roll until another imaginative 3M hero, John Borden, a sales manager, created a dispenser with a built-in blade.

As you can see, members of this company, Richard Drew and John Borden, had learned to build on one another's ideas. The process of innovation is largely incremental. It requires the efforts and contributions of a team if an idea is to be brought successfully to the market place. An idea is rarely marketable in the form in which it is conceived in someone's mind. It generally takes some research, much refining and a lot of hard work – sometimes over years – before it comes into common usage.

With hindsight, in the clear light of success or failure, the good ideas and the less feasible ones may seem obvious. In the early stages, however, the distinction is not so apparent. 'The sublime and the ridiculous are often so nearly related,' wrote Thomas Paine, 'that it is difficult to class them separately. One step above the sublime makes the ridiculous, and one step above the ridiculous makes the sublime again.'

Building on the work of others

A hundred times a day I remind myself that my inner and outer life are based on the labours of other men, living and dead, and that I must exert myself in order to give in the same measure as I have received and am still receiving.
Albert Einstein, *Ideas and Opinions* ed. C. Seelig (Redman, 1954)

The ability to suspend judgement for a time – both as an individual thinker and as a team member – is important. The ability, too, to build on other people's ideas, improving or combining them, is essential. But these two abilities do not exhaust the repertoire of skills required in a member of a truly innovative organization. The ability to criticize in an acceptable and diplomatic manner – in the right way, at the right time and in the right place – also has to be developed.

Constructive criticism

Francis Crick, co-discoverer with James Watson of the double helix, describes in his autobiography *What Mad Pursuit: A Personal View of Scientific Discovery* (Penguin, 1990), a valuable lesson about criticism. He had joined the group studying molecular biology in the Cavendish Laboratory at Cambridge:

> I received another lesson when Perutz described his results to a small group of X-ray crystallographers from different parts of Britain assembled in the Cavendish. After his presentation, Bernal rose to comment on it. I regarded Bernal as a genius. For some reason I had acquired the idea that all geniuses behaved badly. I was therefore surprised to hear him praise Perutz in the most genial way for his courage in undertaking such a difficult and, at that time, unprecedented task and for his thoroughness and persistence in carrying it through. Only then did Bernal venture to express, in the nicest possible way, some reservations he had about the Patterson method and this example of it in particular. I learned that if you have something critical to say about a piece of scientific work, it is better to say it firmly but nicely and to preface it with praise of any good aspects of it. I only wish I had always stuck to this useful rule. Unfortunately, I have sometimes been carried away by my impatience and expressed myself too briskly and in too devastating a manner.

In summary: team creativity cannot be organized, but it can be encouraged. There are structures that foster it, providing you have selected the right participants. The ethos of a group or organization is obviously important. The right climate will encourage people to express ideas, however half-formed. Members are able to discipline themselves in order to suspend judgement. They listen for ideas. They build and improve on one another's contributions. In other words, the *conversation* in that team or organization is positive, confident but realistic, and essentially constructive. There is a musical relationship between the individual thinker and the group. The 'solo' thinker may suggest themes developed by a section of the orchestra; another soloist may take forward a refrain identified by the players as a whole.

COMMUNICATING ABOUT INNOVATION

Remember that progress motivates. If you never give people feedback they will soon lose interest. Good communication supplies a vital dimension in the climate that fosters innovation. It's largely an organization matter, and as such it's covered more fully in the next chapter. But you as an individual manager have a responsibility to ensure that it happens. That means you should:

- Seize opportunities to talk to your people about the importance of new ideas for improving the product range and reducing costs. Give examples and tell stories of changes that have been successfully implemented.
- Explain why suggested ideas have been accepted or rejected for further investigation and development. What are the selection criteria for ideas? Give your team regular progress updates on the passage through the organization of ideas that originated within its discussions.
- Give recognition and reward appropriately – the most powerful communication of all – the ideas that really do make a difference for the better to your business.

Always bear in mind that true communication is a two-way process. You should be a *listening* leader. Everyone that works for you has ideas. Are they listened to?

QUIZ
A staff member working in the transport industry got himself into the *Guinness Book of Records* with the number of suggestions he offered during a career of over forty years. Tick the number of ideas he came up with:
Over 2,000 ☐   Over 20,000 ☐ Over 1,200 ☐   Over 8,000 ☐ Over 6,000 ☐

Answers on page 196.

As your people get the message, and the training you give them in the seven habits of successful creative thinking begins to have its effect, ideas will start popping up all over the place.

In communicating with others your first target as a manager must be the assumptions and fixed ideas of the organization: the luggage it brings from its successful past. Organizations and long-established groups are like individuals in that respect. 'It is not only what we have inherited from our fathers that exists again in us,' wrote playwright Henrik Ibsen, 'but all sorts of old dead ideas and all kinds of old dead beliefs ... They are not actually alive in us; but they are dormant, all the same, and we can never be rid of them.' Well, you must try to rid yourself of these historical handicaps, otherwise your team will never jump the fences that line the course of innovation – as we shall now see.

OVERCOMING OBSTACLES TO INNOVATION

Innovation extends beyond having a good idea. Just imagine for a moment that you are the senior police officer in charge of a large crime detective department of a major police force. A bright, enthusiastic young detective officer asks to see you first thing on Monday morning. 'I saw that Al Pacino film *Sea of Love* on Saturday night,' he begins, 'and next morning, while I was digging up some vegetables, I had this inspiration that we could do something like that here in Sheffield...' What would you say? Here are some phrases that might come to mind (before reading this book, I hasten to say):

'Don't be ridiculous.'
'Interesting, but we've spent our budget this year.'
'It's a good idea in principle, but...'
'It may work in the films, but not in reality.'
'Come, come – will this help my promotion prospects?'
'You're no Al Pacino.'
'It's all right for New York, but it won't work here.'
'We tried something like that eight years ago which was a complete failure.'
'We don't have time for that sort of thing.'
'The chief constable wouldn't approve.'
'The criminals will see through it.'
'Give them free television and video recorders? The media would make us a national laughing stock.'
'Has someone else tried it?'
'Wait until you are a bit more senior before you come up with such hare-brained ideas.'
'Things are bad enough here already without trying out new ideas.'

How would an *equivalent* proposal be received in your organization? Such *idea killers* as those given opposite are the first line of defence against innovation in many organizations. Most ideas die on this sort of verbal barbed wire. But ideas can be killed in many other sorts of ways. They can be drowned in faint praise, poisoned off by the finance director or mangled into nonsense by the computer.

A creative thinker – whether an individual or group – has to be prepared to wade through cold green breakers of initial rejection or criticism in order to swim out into the open sea. As an innovative leader and manager you should be there, alongside, to act as a lifeguard. Your support could make all the difference when the waves stretch endlessly ahead. The conservative forces in organizations really are still surprisingly strong. Learn to delay evaluation – and persuade your managerial colleagues to adopt the same policy. Here is the organizational equivalent of personal Habit Four – Suspend Judgement.

Remember, there is a holistic way of thinking about ideas: as seeds or embryos that grow if they are in the right environment, given the proper nourishment and warmed by the sun. But these seedlings or babies are most vulnerable at the points in time and space nearest to their birth. They are like those baby turtles on Pacific shores that have to scamper down the beach to the sea, braving all the hazards such as the wheeling seagulls in the sky above.

Eventually an idea will grow stronger. Then you can allow it to stand on its own feet and to take its chances in the market place of ideas. Here ideas will expect to be examined, measured, and tested. They are up for sale and will face the usual questions. Do they have any real value? What price is anyone prepared to pay for them? How do they compare to rival ideas that are also bidding for time, money and other resources?

The obstacles in organizations to innovation bear more than a family resemblance to the barriers to creative thinking in individuals listed and described on page 152. That isn't surprising if you accept the analogy of an organization with an individual person, like others in some respects (needs) but different (corporate personality) in others. Pursuing that analogy, some organizations will exhibit far more internal blocks or barriers to innovation than others.

EXERCISE 12: Overcoming obstacles to creativity
1. List *five* obstacles in your organization that you know hold back individuals from coming forward with ideas.
2. Now add *three* examples of when good ideas have fallen at the first hurdle.
3. Next, draw up a brief (one side of A5) Innovation Statement as a guide of best practice for your fellow managers.
4. Circulate it to *ten* of them. See if you can gain their commitment to a version they have helped you to revise by the end of one month from now.

THE CREATIVE LEADER

It is now clear that, apart from being able to provide general direction and perform the necessary leadership functions – defining objectives, planning, controlling, supporting, reviewing – to meet the three overlapping areas of task, team and individual needs, leaders who encourage creativity have some distinctive characteristics:

- *A willingness to accept risk* The potential downside of freedom given to a colleague or team, as we have seen, includes mistakes, failures or financial loss. Delegation should not mean abdication, so you as the leader may well have been a party to the risk. You may at least have understood the potential consequences of things not going as intended or planned. In any managerial role you have to be willing to accept an element of risk. Without freedom there would be no mistakes. But to eliminate freedom is the biggest mistake of all, for freedom alone breeds innovation and entrepreneurial success. Mistakes are a by-product of progress. Learn from them, but do not dwell on them.

- *An ability to work with half-baked ideas* Ideas seldom leap into the world fully-formed and ready to go. They are more like newborn babies, struggling and gasping for life. Creative managers show by example the value of listening to half-developed ideas and building upon them if they have merit. They hesitate before dismissing an ill-formed idea or an imperfect proposal, for it may contain the germ of something really useful. It follows that a creative manager is a *listening* leader.
- *A willingness to bend rules* Rules and systems have their place, but they can obstruct the process of innovation dreadfully. A leader, as a member of the management team, should respect systems, rules and procedures but should not *think* like a bureaucrat. Sometimes creative dyslexia – the inability to read rules – is a strength rather than a weakness. Rules can sometimes be stretched where they cannot be broken. Otherwise you end up by being bogged down in organizational treacle – or, as Charles Dickens said, 'Skewered through and through with office-pens and bound hand-to-foot with red tape.' Remember that Nelson once put his telescope to his blind eye. Having a blind eye can be a strength on occasion, not a weakness.
- *An ability to respond quickly* Continuing with the newborn baby analogy, some new ideas or projects need

sustenance quickly if they are going to survive. Leaders who induce creativity should have a flair for spotting potential winners. But that is not enough. The innovative organization must also have leaders who are able to commit resources without having to defer everything to committees or upwards to Higher Authority. Being able to allocate or obtain small resources now may be far better than being able to summon mighty resources in a year's time when it is too late.

- **Personal enthusiasm** Only leaders who are highly motivated themselves will motivate others. Enthusiasm is contagious. Moreover, enthusiastic leaders and colleagues tend to be intellectually stimulating ones. 'Man never rises to great truths without enthusiasm,' wrote Vauvenargues.

KEY POINTS: HOW TO MANAGE INNOVATION

- A house is made up of individual bricks. The quality of an innovative organization depends ultimately and largely upon the quality of the people you employ. Machines do not have new ideas. Computers cannot create. Money alone cannot create a satisfied customer.
- Look out for the twelve characteristics – or clusters of characteristics – that mark creative people. Ensure that many of those you appoint have some of these characteristics.
- At the core of team creativity is the capacity to build upon or improve other people's ideas, and to subject your own ideas to the same process. 'The typical eye sees the ten per cent bad of an idea,' writes Charles F. Kettering, 'and overlooks the ninety per cent good.'
- Building on ideas sounds a simple recipe, and so it is. But

it presupposes a positive and constructive ethos, mutual encouragement, and the ability to listen.
- Although it may be focused in particular meetings or even in departments, such as research and development, team creativity should embrace the whole organization. It should be a basic theme in the endless conversation of any organization that seeks to be innovative.
- Ideas leading to innovation are more likely to come from people when their leaders expect them.

> *More creativity is the only way to make tomorrow better than today.*
>
> Anon

CHECKLIST:
HOW TO MANAGE INNOVATION

Answer Yes or No to the following:	Yes	No
When you interview external candidates for jobs do you take into account their relative interest in *improving* the job as well as *doing* the job?	☐	☐
Have you a clear idea of the main characteristics of creative people?	☐	☐
Is your organization able to tolerate or accumulate the downsides of the creative personality in order to benefit from fresh thinking and new ideas?	☐	☐
Does your team build on each other's ideas, using them like stepping stones towards solutions?	☐	☐
Has everyone who works in your organization had a minimum of one day's training in creative thinking techniques within the last twenty-four months?	☐	☐
Have you given your own team any training in creativity and innovation?	☐	☐

List your strengths and weaknesses as a creative manager:

	Strengths	*Weaknesses*
1.	....	....
2.	....	....
3.	....	....
4.	....	....

11

THE INNOVATIVE ORGANIZATION

'The creative act thrives in an environment of mutual stimulation, feedback and constructive criticism – in a community of creativity.'
William T. Brady

The third dimension on that oyster-shell model of innovation you saw on the first page of this book, is the **organization.** What makes some organizations more innovative than others? Why do some organizations act as magnets to people with flair, imagination and creative energy, while others, by contrast, cannot find large enough bribes to secure or keep talented staff?

'Innovation is our motto,' one senior manager told me. Then he added with a shrug of his shoulders: 'Our trouble is that we don't practise it.' He had shown me all the paperwork: the Chief Executive's Vision, the Company Mission, the Core Values of the Corporate Philosophy. Yes, I had counted at least nine mentions of innovation and another five references to the importance of creative thinking. Impressive documents, written by professionals. 'All window dressing,' said my companion. He continued: 'Our internal innovative capability is severely limited by our constant quest for short-term profits. They can measure those! But they can't measure the most important resource of all – our innovative power. So people have just switched off.'

It is one thing to have a vision of an organization standing tiptoe with innovative zeal; it is quite another to realize that vision. That is why many organizations so often find themselves on that journey homeward to their old habitual self. The excitement of the latest change programme wears off, the old patterns break through the thin coating of concrete ... and the regression is under way.

Setting aside what organizations *say* about innovation, are there any characteristics that mark out those who actually practise it? As you might guess, there are. Here are five key ingredients, which overlap considerably, that can be clearly identified and described:

- Visible commitment at top level
- A climate that encourages teamwork and innovation
- A toleration for failure to balance risk-taking
- Open and constructive communications
- Flexibility in organizational structure

Together these factors inoculate your organization against the fate of the dinosaurs. For 'an established company which in an age demanding innovation is not capable of innovation is doomed to decline and extinction,' predicted management sage Peter F. Drucker many years ago. 'And a management which in such a period does not know how to manage innovation is incompetent and unequal to its task. Managing innovation will increasingly become a challenge to management, and especially to top management, and a touchstone of its competence.'

TOP LEVEL COMMITMENT

The top leadership team – the chief executive and executive directors – need to show visibly and audibly that they are committed to positive innovation. Their weight and influence is necessary to overcome the barriers and resistances to useful change that innovators often encounter. For the process of innovation may become too slow if vested interests are allowed their head. What may seem a corporate opportunity to you may be perceived by others as a departmental threat. It is your job as a leader at any level in the organization to facilitate desirable change and to encourage that attitude throughout the management team.

Without real commitment from the top, real innovation will be defeated again and again by the policies, procedures and rituals of almost any large organization. Gone are the days when chairmen or managing directors could afford to wait for others to push creative change and innovation in their organizations. Now they have to throw their weight into the scales that favour positive, proactive change. To change things calls for leadership; to change them before anyone else is innovation.

Abraham Lincoln

Genuine leaders, such as Abraham Lincoln, are not only instruments *of* change they are catalysts *for* change ... Lincoln effected the change needed by being extraordinarily decisive and by creating an atmosphere of entrepreneurship that fostered innovative techniques.

Years before assuming the presidency, Lincoln had shown his interest in innovation when, on 10 March 1849 (at the age of 40), he received a patent for making grounded boats more buoyant (he is the only United States president to have secured a patent).

After he became president and moved to Washington, Lincoln took a keen interest in any technical development, especially in weapons, that could hasten the end of the Civil War. He attended demonstrations, talked to inventors and personally tested new weapons, such as the new breech-loading and repeating rifles.

Overall, Lincoln's philosophy and handling of the most up-to-date technology of the time was brilliant ... He realized that, as an executive leader, it was his responsibility to create the climate of risk-free entrepreneurship necessary to foster effective innovation.

> Donald T. Phillips, *Lincoln on Leadership*
> (Warner Books, USA, 1992)

Senior executives who believe in innovation will make the necessary resources available. Every organization should have a development fund for helping promising ideas in those vital early days. Quite small sums can be remarkably effective, not least in encouraging the team or individual responsible for the innovation-in-embryo. Later more substantial funding needs to be available for start-up situations.

For organizations are unlikely to be innovative – to introduce change and make it effective – if they lack a sense of direction. If you are not facing the future and wanting to move forward, why change? But do you as a chief executive – or your chief executive – see the implications for your role?

CASE STUDY: SOICHIRO HONDA – A LEADER FOR INNOVATION

At 21 years of age Soichiro Honda opened his own repair shop. He then moved into manufacture, starting with an initially unsuccessful venture in making piston rings – unsuccessful because he lacked the necessary knowledge. Meanwhile he completed a racing car with his brother Benjiro. Both of them drove the car in its first race, which ended abruptly with a crash and some serious injuries.

Helped by the practical experience acquired during these years, now supplemented by a university course in engineering, Honda began to invent and innovate. One of his early patents was for a piston ring polishing machine, revolutionary in design and simple to operate. After three long years of trial and error in making piston rings his persistence had paid off and he produced some excellent ones. 'Those days were the most difficult times with many hardships,' he later recalled.

With the end of the Second World War, the company that Honda had established ceased to be directly involved in pistons. The genesis of a new direction lay in an apparently chance event. One day a friend brought him a 50cc engine that had been used by the Japanese army. He wondered if Soichiro could find a use for it. In those austere post-war years the transport system in Japan was very poor. The trains and buses were so crowded with people that they sometimes had to use the windows for exits.

When Soichiro looked at the engine he thought to himself, 'A bicycle with a power source would be great.' The problem, however, was that everything was in such short supply that it would be impossible to obtain fuel tanks. At this point Soichiro's creative mind took over. What could he use as a fuel tank? He thought of everything and finally came up with the idea of using a hot water bottle. He attached one to the bicycle and began experimenting. This was the beginning of the motorized 'Putt-Putt' bike, named to mimic the sound it made.

In that post-war period it was very difficult to get petrol, so Honda had to rely on the oil that could be extracted from the roots of pine trees. After repeated trials he finally made a motorized bicycle. Since there were only 500 of the old Japanese engines available, Soichiro's next step – a decisive one – was to start building his own. And so the Honda Motor Company was born, with a capital of one million yen.

Within the next few years Honda's motorcycles were winning the main prizes on the world's racing circuits. But Soichiro never lost sight of the memory of one of the early automobiles that had roared through his boyhood village where his father worked as the local blacksmith. Using Formula 1 racing as his testing ground he moved into car production for the mass market.

Even then the environmental problems caused by gas emissions were becoming an issue. But Honda saw it as an opportunity, not a problem. The man who would become the fourth president of Honda, Nobuhiko Kawamoto, remembered those days: 'When the Clean Air Act of the United States of America was passed by the US congress in 1972, Soichiro said that even though Honda was behind General Motors and Toyota in sales and car production experience, we were all equal at the starting line in respect of the development of the low emission engine. Soichiro inspired us to believe that we were equal with GM and the other bit automakers.'

In order to respond to the needs of society, Soichiro turned to the ability of his young engineers. He believed in the power and competence of young people. Headed by

Tadashi Kume (who later became the third president of Honda), the Honda engineers successfully developed a low emission water-cooled engine, the CVCC. The Civic, equipped with this engine, became a worldwide hit. The United States in particular hailed the Honda engineers' ingenuity.

Satisfied that his spirit of thinking ahead of the competition rather than imitating others had permeated all of his employees, Soichiro decided to retire from his position as president of the company. Upon his retirement in 1973, he became the top advisor to the company and its young engineers.

Kiyoshi Kawashima, the second president of Honda, said: 'Soichiro did not like to copy what other people did. He used his own ideas and creativity to make things. A man with dreams, who helped his employees fulfil their own dreams. I learnt from him that a company without a dream will lose its place in society.'

The fourth Honda president, Nobuhiko Kawamoto, recalled: 'Soichiro was characterized by his desire for exploration and his spirit to be true to himself. Although he ordered us to stop the Formula 1 project, he continued to be interested in racing until his death. Even a few days before his passing, he became very animated and was very interested in discussing Formula 1 with me.'

Even after he accomplished his goal, Soichiro never stopped moving forward. The curiosity and spirit that made the young boy run after that noisy, speeding vehicle on that dusty country road so long ago, never left him. That oil-spotted track left in the road was always with him. His legacy to all who knew him, and even those who didn't, is sometimes referred to as 'Hondaism', the spirit to praise the dynamic force of youth and encourage young creative minds.

Remember, it doesn't all depend on the chief executive – important though that strategic leadership role is. What counts is the whole leadership team, including all the operational or business leaders and all the team leaders. In organizations operating towards the creative end of the spectrum it is essential that all managers should be directors or leaders, otherwise they will lack credibility. Leadership implies leading by example, which usually means, in this context, having a contribution to the innovative process yourself.

A CLIMATE THAT ENCOURAGES TEAMWORK AND INNOVATION

The importance of a culture, climate, atmosphere or ethos favourable to creativity is widely recognized. For in innovation *none of us is as good as all of us*. The aim is to develop a community of creativity as defined by William T. Brady at the head of this chapter. Those three elements – mutual stimulation, feedback and constructive criticism – take us right to the heart of the matter. Every leader, including you, can help to create that climate as a by-product of practising the seven habits of creative thinking and acting as a role model in the management of innovation. If you are not part of the solution, you are part of the problem.

The organizational climate should be an open one that encourages participation. There must also be a willingness to provide the relevant facts and information to enable employees to make an informed contribution. There should be a long-term commitment to the positive management of change on the part of managers at all levels, together with a readiness to provide the necessary resources for education and training.

In a corporate culture that favours new ideas and innovation, the difficulty is, of course, is how to combine these ingredients with the high degree of structure, discipline and routine that is required to manufacture products and deliver a proper customer service. Not all members of the organizational team will be equally capable in both these aspects. But then the essential characteristic of a team, as we have seen, is that it is composed of people with *complementary* temperaments, sets of qualities, interests, knowledge, and skills.

Notice that teamwork at this level calls for excellent lateral communication and flexibilities of structure – two other core characteristics of innovative organizations.

One implication will be that individuals who show promise as creative or innovative thinkers will spend some of their time outside their departments or divisions contributing as members to intellectual project groups. Set up to solve macro-problems or to explore strategic opportunities, these task teams will be inter-disciplinary in nature, for diversity breeds creativity.

A TOLERATION FOR FAILURE

As Nobel Prize winner Sydney Brenner said, 'Innovation is a gamble.' If you have never worked on the edge of failure, you will not have worked on the edge of real success. Creative people respond well to an organization that encourages them to take calculated risks.

It is virtually impossible to innovate without accepting an element of risk. You can and should calculate risk and adjust your exposure to match your resources. But you cannot eliminate risk altogether and still see yourself or your organization as being creative and innovative. 'Nothing ventured, nothing gained.'

> *The most important of my discoveries have been suggested to me by failures.*
>
> Sir Humphry Davy

The downsides of risk are mistakes and failure. In any entrepreneurial and innovative enterprise there will be such failures. They are, of course, quite different from the failures that arise from indecision and inaction. Business leaders must accept this downside and pick up the bill bravely. The possibility of failure should not be used as an excuse – it often happens – to pull in the horns of creative thinking and innovation.

There should be a discussion after each failure – in order to learn the lessons, not to dole out the punishments. You will usually find that there were warning signs of impending failure that were ignored. One important lesson to be learnt from such discussions is that managers should face the unpleasant task of ending potential failures before they gather too much momentum.

'There are risks and costs to a programme of action,' said John F. Kennedy. 'But they are far less than the long-range risks and costs of comfortable inaction.' In other words, if you take risks you may make mistakes, but if you do not take risks you are doomed to failure.

The former president of 3M, Lewis W. Lehr, had some wise things to say about the need to accept mistakes – but only if they are first-time ones. The corporate culture of 3M has a clear policy or tradition on the matter:

> The cost of failure is a major concern for innovators – since that is what will happen to most of them at one time or another. We estimate at 3M that about sixty per cent of our formal new-product programmes never make it. When this happens, the important thing is not to crucify

the people involved. They should know that their jobs are not in jeopardy if they fail. Otherwise, too many would-be innovators will give in to the quite natural temptation to play it safe. Few things will choke innovation more quickly than the threat of losing a job if you fail.

We have a tradition of accepting honest mistakes and failures without harsh penalties. We see mistakes as a normal part of business and an essential by-product of innovation. But we expect our mistakes to have originality. We can afford almost any mistake once. Those who choose to lead high-risk, new-product programmes know that their employment will not be threatened. This attitude of management eliminates one of the major barriers to innovation in large companies.

As any business grows, it becomes necessary to delegate responsibility and to encourage people to use their initiative. That means allowing them to do their own jobs in their own way. If the person is essentially right, the mistakes that he or she makes will not be as serious as the greater mistake of trying to specify in an autocratic way how everything should be done or to insist on all decisions being made at head office. A top management that is destructively critical when mistakes occur will smother initiative and enterprise. When that happens – goodbye profitable growth.

Have you made enough mistakes yet?

In 1943, Wernher von Braun was working on a rocket the Germans hoped would destroy London and end the war.

Producing this rocket required new metals, new fuel, new guidance systems, new everything. Von Braun's superiors were impatient to move the project to completion. They were angered by the many changes he had sent to the factories responsible for manufacturing the rocket.

'You are supposed to be the ultimate brain in this operation ... do you know off-hand how many last-minute changes you've made in your rocket plans ... since you started two years ago?'

They waved a piece of paper before von Braun. 'Make a guess, Professor. How many changes have you sent to the factories?' And there the ridiculous figure was: 65,121. It was accurate. Von Braun acknowledged his 65,121 mistakes.

He then estimated that he would make 5,000 more before the rocket was ready. 'It takes sixty-five thousand errors before you're qualified to make a rocket,' he said. 'Russia has made maybe thirty thousand of them by now. America hasn't made any.'

In the second half of World War II, Germany alone pounded her enemies with ballistic missiles; no other country had them. And when the war was over, Wernher von Braun became the 'ultimate brain' in America's space programme. Only a few years – and many mistakes – later, America put a man on the moon.

> Based on the account in James Michener, *Space*
> (Secker & Warburg, 1982)

It is essential for business leaders, then, to accept the risk element in decision making, especially when it comes to innovation. Risk means the possibility of loss or injury. But if you never go out on a limb you will not pluck the best fruit.

A chief executive of a company was recently summoned to the corporate headquarters of the international group he had worked for. He had just made a substantial loss on a major project and was therefore expecting to be dismissed. At the end of his meeting with the president, however, neither the loss nor his imminent departure had been mentioned. As he stood up to leave he said, 'It's good to know

I have still got a job. I must confess I thought you would fire me today as a result of that substantial loss.'

'Fire you?' replied the president. 'Hell, no, your education has just cost me one million dollars!'

The 147/805 rule

You often hear the expression, 'We have tried that several times and it doesn't work here.' But next time you step into an aircraft remember that the Wright brothers tried 805 times before they achieved sustained flight.

Edison, for his part, failed 147 times before he hit upon the solution to the electric light bulb. What separates an idea from success is often perseverance.

Provided that failure is not the consequence of recklessness or incompetence, innovative organizations will not exact revenge or make scapegoats. It is usually easy to be wise after the event. Although you should endeavour to be wise before the decision, it is no good belabouring yourself for not knowing then what you know now. Put it down to experience in your ledger of success and failures. As they say, you can't win them all. Oscar Wilde once defined experience as the name we give to our mistakes.

It all comes back to the real commitment and leadership of the chief executive and the top management team. If they are firmly resolved upon profitable growth through team creativity, then the challenge of innovation will be met. Even with a good track record, do not leave anything to chance. Certainly, the best way to lose an innovative edge is to spend too much time admiring a successful past. A good reputation is history, nothing more. Good companies must always search for excellence.

TAKING A LONG-TERM PERSPECTIVE

The criterion of short-term profit – the bottom line each quarter – is clearly inappropriate when it comes to developing and introducing new products and services. 'No great thing is created suddenly,' wrote the Roman philosopher Epictetus, 'any more than a bunch of grapes or a fig. If you tell me that you desire a fig, I answer you that there must be time. Let it first blossom, then bear fruit, then ripen.' So it is with any commercially viable new product or service.

In comparison with Japan, for example, where banks and corporations take a more long-term view, Western financial institutions and shareholders in countries such as Britain and the United States are notorious for their *short-termism*. Such thinking and policy making cannot encourage industry to innovate. Banks in particular in these threatened countries need to recover a larger purpose than mere short-term profit. For they exist in part to provide a service to business and industry, who are the engine-room of their economies. Too often they fail to do so. At least they should now adjust their sights to take a more medium-term view – the good old British compromise – when deciding on the investments they should make for the future.

British entrepreneurs such as Richard Branson and Andrew Lloyd Webber are among those who have bought their public companies back into private ownership. They have resented what they see as too much emphasis on producing short-term profits at the expense of long-term growth. 'Being private enables us to adopt the Japanese approach of building market share slowly and then waiting for profits,' says Branson. 'Most of the year, running a public company, was spent worrying about next year's profits. Since going private, I haven't once asked for a profit forecast.'

Innovation should not be a reactive process but part of a long-term strategy that gives direction. It needs to be fed by the dynamo of a corporate sense of purpose. Such a strategy will balance the present needs of producing and marketing *existing* goods and services – the commercial priority – with the middle- and long-term requirement of *research and development*. A balanced and coherent strategy will enable your organization to build on its past successes and create its desired future. *It is the only sure pathway to profitable growth.*

There are always reasons for not becoming an innovative organization, not least the fact that it costs money to go down that path. But can you afford the cost of the alternative?

KEY POINTS: THE INNOVATIVE ORGANIZATION

- Without a leadership team at the top who value product quality, new ideas and innovation, and who constantly struggle to keep organizations moving towards these guiding stars, there will be no sustained and profitable growth.
- 'If the trumpet give an uncertain sound, who shall prepare himself for the battle?' The top management team should seek ways of making their commitment to positive and useful change visible to all concerned.
- If an army marches on its stomach then a business marches on its investments. Research and development is the seedcorn for future innovation. It is not a cost but an investment, one with no predictable outcome. Is your organization making that investment?
- Flexibility is the ability – personal as well as corporate – to modify, alter and perhaps radically change what you are doing. Rigid or inflexible structures produce inertia.
- Risk is the companion of innovation. As the Japanese

proverb says, 'Unless you enter the tiger's den you cannot take the cubs.'
- The innovative organization is the reverse image of bureaucracy: flat rather than pyramidical; decentralized decision-making and devolved responsibility; informal instead of formal; emphasis on lateral as well as vertical interaction; rules kept to a minimum; and positive about appropriate and properly calculated risks.
- Wise strategic leaders like Soichiro Honda always stay in touch with the young in their organizations, for youth, courage to think new ideas and boldness to take risks tend to go together. As Francis Bacon put it: 'Men of age object too much, consult too long, adventure too little, repent too soon.'

Man is pre-eminently a creative animal, predestined to strive consciously for an object and to engage in engineering – that is, incessantly and eternally to make new roads, wherever they may lead.

Dostoevsky

CHECKLIST:
THE INNOVATIVE ORGANIZATION

These 20 questions will give you a rating on where your orgainization stands in terms of innovation:

1 = poor 2 = average 3 = good 4 = very good 5 = excellent

Rate your top executive management team's commitment to innovation	1 2 3 4 5
How far does the organization's vision emphasize the need for Innovation?	1 2 3 4 5
How well is that vision or philosophy communicated to all?	1 2 3 4 5
Is your chief executive an enthusiastic leader of change?	1 2 3 4 5
What is the level of mutual stimulation, feedback and constructive criticism?	1 2 3 4 5
Assess the organization in terms of its internal teamwork.	1 2 3 4 5
Is regular and effective use made of project teams?	1 2 3 4 5
Are failures and errors accepted as part of the psychological contract in exchange for proper risk-taking?	1 2 3 4 5
What's your record for retaining creative and talented young people?	1 2 3 4 5
Are rewards, promotion or advancement linked at least in part to innovation?	1 2 3 4 5
Evaluate the state of lateral communications.	1 2 3 4 5
Are there plenty of informal opportunities for exchanging ideas?	1 2 3 4 5
Rate your organization's freedom from 'if only' excuses.	1 2 3 4 5

Are resources made available to support new initiatives?	1 2 3 4 5
Assess the structural flexibility in the organization as a whole.	1 2 3 4 5
Are decisions really pushed down to the lowest point at which they can be taken?	1 2 3 4 5
Does everyone in the organization see themselves as involved in the innovative process?	1 2 3 4 5
Has the organization adopted a long-term perspective over innovation?	1 2 3 4 5
Is innovation part of an overall strategy for creating tomorrow's organization out of today's one?	1 2 3 4 5
Is it much fun to work in your organization?	1 2 3 4 5
Totals	

How to Interpret Your Score

70–100   Congratulations – you work in a highly innovative organization with a bright future.

40–69   Show some moral courage. Photocopy this checklist and take it to your chief executive as a means of raising the issue for urgent discussions. Remember to take with you some positive suggestions.

10–39   Drastic action is needed if this organization wants to stay in business. In the meantime, get your parachute on.

12

THE ART OF BRAINSTORMING

Linus Pauling, the Nobel Prize-winning scientist, once said, 'The best way to get good ideas is to have lots of ideas.' That is as true for a team or an organization as it is for an individual. But how do you facilitate this productivity of ideas as a manager?

Unlocking the safe in which other people keep their ideas is sometimes easy. People love to talk about their work – especially if they love their work. You only have to show a certain interest. Ask a few searching questions and the safe door swings open. Or, to use an analogy from my days on Arctic trawlers, you only have to pull the knot at the cod-end of the net and a mass of silver fish of all shapes and sizes cascades onto the deck.

There are times, however, when you may need a burglar's tool-kit for picking the safe of a team's inner mental resources. In this chapter we shall look at the best known and most often used core technique for generating ideas – brainstorming. Look on it as a tool that can help your business to survive and grow by unlocking doors to ideas that are already there. Incidentally, the following discussion does contain some reminders and developments of those seven personal habits of successful creative thinkers outlined in Part Two. For teams or groups need their own counterparts to the seven habits.

BRAINSTORMING DEFINED

Brainstorming has been defined as a method of *getting a large number of ideas from a group of people in a short time.* It is essentially a group activity that follows a relatively formal procedure to generate as many ideas as possible without stopping to evaluate them.

The origins of brainstorming go back to Alexander Osborn, the founder member of a large advertising company, whose book *Applied Imagination* first appeared in 1953, and has served as a primary textbook for brainstormers ever since. According to this book the first brainstorming sessions were held in America in 1939. Undoubtedly the best known approach to producing ideas, the technique remains a key part of 'quality circles' and 'total quality management' programmes today. Introduced to improve quality of products or services, these programmes almost always hinge on small groups or teams seeking new ideas, both on product improvements and issues of efficiency, such as reducing costs.

The fact that brainstorming has been known and used for some time and has now proved so relevant in total quality management should serve as an encouragement to take it seriously. 'Truth is the daughter of Time,' says one proverb, while another declares that 'Time tries Truth'. In management, though, I suggest that if a model, technique or concept survives the vagaries of fashion for more than ten years it merits your deeper attention! Beware of the 'newer-is-truer' heresy. We need novelty – new ideas or fresh ways of expressing old truths – to stimulate our minds, I hasten to add. But action is best based upon rock that has the gold streaks of truth in it. There is such a thing as fool's gold. Brainstorming, however, is both well-tested and tried. It's also simple, which is no small advantage. It's soundly based on one of the habits of successful creative thinkers – suspending judgement – though a good brainstorming group will touch all the other chords or notes as well.

BRAINSTORMING TECHNIQUES

The word 'brainstorming' has now become synonymous with any kind of ideas generation – much as 'hoover' is to vacuum cleaner – and many variations and supplementary techniques have emerged. I would suggest, however, that you begin by focusing on the principles of the original technique – and your own skills as a brainstormer – before moving on to its various derivatives.

In the following pages you will see that for the most part I have assumed that you are a team leader in the brainstorming session. But of course, you can put your skills to work equally well if you are a team member.

It may incidentally be worth giving out this chapter to your team as an *aide-mémoire* of the useful techniques available to them. The team as a whole, or an individual member, may see a need or an opportunity for using a particular technique that you as leader might miss.

BRAINSTORMING IN PERSPECTIVE

Does brainstorming really work? There are relatively few objective evaluation studies on the technique. According to one such study the quantity and quality of solutions increased in brainstorming groups when compared to those that allowed unrestrained critical evaluation of ideas without following brainstorming rules. Yet another researcher found that both kinds of groups produced ideas of the highest quality: brainstormers merely threw up more low-quality ideas. Originality did not come into the definition of quality in either of these pieces of research.

A more important study compared individual, as opposed to group, brainstorming and offered the conclusion that although a group can produce more ideas than one individual, a number of individuals brainstorming alone can throw up more ideas than the same number working as a group. Moreover, those individuals working on their own produce ideas of a higher average quality, suggesting that 'a person alone is still freer in thought than one in a brainstorming group, despite the suspend-judgement, no-criticism rule'.

Clearly one question is, how far does quantity of ideas in itself lead to right choices? Perhaps a brainstorming group can be compared to a computer that can produce many more possibilities for moves than chess grand masters, but lacks the human ability to single out without extensive analysis the more feasible or possible move.

The evaluation evidence from research on brainstorming, however, is inconclusive. Certainly we can say that brainstorming stimulates a free flow of ideas, but thereafter the group dimension may become less important. In other words, group brainstorming may be a good way of encouraging creative thinking in individuals who may later work on their own or in more informed situations with others.

With regard to the latter, in some ways brainstorming is an artificial stimulation of what occurs when creative people talk to each other. Sydney Brenner mentioned sharing a room for twenty years with Francis Crick, one of the co-discoverers of DNA. 'At least two hours a day we talked nonsense about anything.'

The research also suggests that brainstorming works best on specific and limited but open-ended problems. Given these criteria and the valuable suspension of judgement, it is a most useful technique for stimulating a quantity of ideas. Matters such as product improvement, developing new advertisements, or thinking up new brand names lend themselves to brainstorming. For original creative work, which requires extensive preparatory work, depth mind activity and the guidance-system of values, it is much less effective. It is positively harmful as a philosophy: namely that 'group-think' can replace individual creativity.

Brainstorming, then, is a useful technique for releasing ideas by overcoming inhibitions, cross-fertilizing minds and getting away from fixed rules or hidden assumptions. It needs to be prepared and planned, and followed up with evaluation.

Use it selectively, where you have some definite but open-ended problem areas. It won't solve all the problems but it can help you to by-pass some of the traffic jams that often build up on the traditional routes to problem solving.

Remember that however creative the group feels it has been, what you finally decide on has to work. Brainstorming and other techniques for increasing creativity will help you to break new ground, but eventually you will bring into play the meta-functions of analysing and valuing. For clear thinking about the pros and cons of the preferred solution and good judgement are required before you start to make it happen.

THE GROUND RULES

The simple basis of brainstorming is to allow ideas to flow freely from a group without criticisms being voiced. The rules as laid down by Osborn have not varied substantially since 1939, and are described in the table over the page.

FOUR RULES FOR BRAINSTORMING	
ROLE	NOTES
Suspend judgement	Criticism is ruled out. Adverse judgement of ideas must be withheld until later. Do not evaluate.
Free-wheel	Free wheeling is welcomed. The wilder the idea, the better; it is easier to tame down than to think up. Let your *mind drift*.
Strive for quantity	Quantity is the aim. The greater the number of ideas, the more the likelihood of success. Aim at, say, a hundred ideas in a period of fifteen to thirty minutes.
Combine and improve	Combination and improvement are sought. In addition to contributing ideas of their own, participants should suggest how ideas of others can be turned into better ideas; or how two or more ideas can be joined into still another idea. Hitch hike on other people's ideas.

Brainstorming, as I have mentioned, is based upon *the principle of deferred judgement*, or, as expressed by its originator, Alex Osborn, the principle of suspended judgement. The basis is a deliberate alternation of the thought processes. In other words, you should turn on your valuing mind at one time and your creative mind at another, instead of trying to think both critically and imaginatively at the same time.

It's always best not to judge other people's ideas too soon or critically. An angry banker once told Thomas Edison to 'get that toy out of my office!', so Edison took his invention (the phonograph) somewhere else.

Another way of defining these rules might be as follows:

Do not evaluate – the most important rule in brainstorming and the most difficult to observe. Criticism of new ideas is deadly. Not evaluating/criticizing an idea is not equivalent to agreeing that the idea is feasible.

Think wild – the need to create a supportive environment for *new* ideas. It's important that everyone feels free to be way out, far-fetched, hare-brained, stupid, crazy, wrong or off-the-wall. The result of thinking wild is to switch the team's train of thought into new directions.

Think prolific – or generate as many ideas as you can – a way of scanning the field for all the possibilities, whether viable or not. It builds the habit of never stopping at the first idea you have. For in creative thinking, as we have seen, the good idea is often the enemy of the best idea.

Build and bounce – an invitation to everyone to make a conscious effort to build on other people's ideas or bounce off in new directions. It leads to discovering directions that wouldn't have emerged if you brainstormed the problem on your own. The secret of successful brainstorming is to listen to others in order to form a synergy.

EXERCISE 13: Making the best use of brainstorming
1. Select twenty problems in your organization or area of responsibility that would respond to the group brainstorming approach.
2. Select three of them for further work.

You may like to extend the above exercise by assembling a group of between six and ten people for an experiment. Some members should have direct experience or first-hand knowledge of your short-listed problems. Others should be selected so that different perspectives are brought to bear on the problems you have chosen.

Of course, brainstorming isn't necessarily the sole activity of a meeting. It doesn't require a dedicated group meeting specifically and only to brainstorm. The brainstorming technique, for example, is a most useful spanner in the toolbox of any kind of project group. As a project group manager you may suggest its use if the team's paths to solutions or courses of action appear to be blocked, or if you suspect that easy or average solutions are being opted for.

LEADING A BRAINSTORMING SESSION

To be an effective innovator you need to be able to lead a brainstorming session. Good leadership is a critical factor in the success of creative thinking groups. As Osborn pointed out, 'Fiascos are usually due to failure of leadership.' Follow the guidelines in the table opposite.

These are the guidelines. If you are leading the group, however, you may want to put them into your own words since a brainstorm session should always be kept as informal as possible. Here's how one leader interpreted the first principle to one of his groups:

> If you try to get hot and cold water out of the same tap at the same time, you will get only tepid water. And if you try to criticize *and* create at the same time, you can't turn on either *cold* enough criticism or *hot* enough ideas. So let's stick solely to *ideas* – let's cut out *all* criticism during this session.

A few incurable critics may still ignore the guidelines and belittle what others have suggested. Such transgressions should be gently warned against, and – if persistent – firmly checked. For the *spirit* of a brainstorm session can make or break it. Self-encouragement and mutual encouragement are both needed. The kind of criticism that cramps imagination, however, breeds discouragement.

STEPS IN LEADING A BRAINSTORMING SESSION	
Introduce	State the aim and define the four ground rules. Put them on a flipchart or overhead slide. Identify the person who will act as note-taker.
Warm up	If necessary familiarize the group with the procedure by a practice exercise. For example, ask them to 'suggest thirty new uses for any object, such as a brick or paperclip'.
State the problem	Avoid making it either too specific or too general. If it concerns a product, have examples of it there.
Guide	Give people two minutes to think about it and write their ideas down. 'In how many ways can we...' Encourage everyone to contribute as the person who is listing the ideas writes them on a whiteboard or flipchart. Quell any attempts to comment on, criticize or evaluate suggestions. Remind the group to go for quantity not quality – the more ideas the better. Use short questions only for clarification – no lengthy discussion allowed. Discussion and questions should be unnecessary. Control and limit discussion and sustain the free flow of ideas.

If the group falls silent during brainstorming, allow the silence to continue for a full two minutes before contributing. This procedure maintains time pressure as well as giving an opportunity for the individual's depth mind to work.

There are few people who have participated in brainstorming sessions who have not experienced 'chain-reaction' – when minds are really warmed up, and a spark from one mind will light up a lot of ideas in others like a string of firecrackers. Association of ideas comes into play, so that an idea put into words stirs your imagination towards another idea, while at the same time stimulating associative connections in other people's minds, often at a subconscious level.

Putting ideas into words, however ill-informed, is the vital step in brainstorming. As the seventeenth-century poet Edward Young wrote, 'Thoughts shut up want air, and spoil like bales unopen'd to the sky.'

As a slight variation to giving people some time to think individually, try issuing everyone with a pad of, say, twenty 8 x 14 centimetre slips of paper or cards. Having presented the ground rules and the problem in 'how-to' form, ask the group to write down as many answers as they can within five to ten minutes, each one on a separate slip. These ideas can then be used as triggers for a brainstorming session. Or you can shuffle the pack and give the cards out again, asking each person to build four more ideas on the ones written on the cards they receive.

A more substantial variation is to introduce a *matrix* in one form or another in order to stretch people's minds. You can fill in the left-hand vertical column with questions and the upper horizontal one with your areas of interest, such as products, services or costs. Opposite is an example just to remind you of the idea of using matrices in this way:

In the example of a matrix opposite you can introduce another creative dimension (Habit Two: Welcoming Chance Intrusions) by playing the game of Forced Connections, as follows. Supposing the product you are thinking about is a wash-basin. Get the team to wander about the place for ten minutes picking up surrounding objects and make forced connections between them and the product. A daffodil, for example, may suggest a new design of tap. Or an inkwell may suggest a recess cavity for sticking toothbrushes in.

Again, it's worth emphasizing that the benefit of introducing such variations is not only that you will keep the

	PRODUCT A	**PRODUCT B**	**PRODUCT C**
What if ... it became bigger? (thicker, heavier, stronger)			
What if ... it was reduced in size? (thinner, lighter, shorter)			
What if ... it was made or done faster?			
What else could it be used for (without any changes)?			

Idea Development Matrix

team fresh. By so doing you'll also be programming everyone's depth minds. Once encouraged, they will begin to see connections between things apparently disconnected.

Matrices are especially valuable for identifying blank spaces on the creative map. If, for example, you listed your company's *capabilities* in the vertical column and the actual or potential *markets* for them in the horizontal one, you might be able to identify a market area where a particular competence is not being put to work. You can then target that area for brainstorming with an array of what-if, how-to, why-not, when-to and with-whom questions.

Be creative in your use of matrices. For example, you can make a Features Matrix, putting PRODUCTS down one axis and APPLICATIONS or CUSTOMERS on the horizontal one. Try listing in the boxes some of those 'forced' but feasible ideas as you try to 'connect' the two categories in your mind.

THE IMPORTANCE OF FOLLOW-UP

It has been demonstrated that no more than forty minutes should be allocated to the actual session but the participants are usually asked to go on considering the problem and send in further suggestions. These are added to the list already obtained and all ideas are classified into logical categories by the leader. These classified categories are handed to the person who initially submitted the problem. He then evaluates the list, possibly processing ideas by combination, elaboration or additions of his own.

Evaluation may be best done not by the brainstorming group itself but by a small group of five members directly concerned with the problem. If this is the case, keep the brainstorming group informed of the result, otherwise they may want to be excused the next time they are asked! The steps of evaluation are:

- Decide on appropriate criteria
- Pick out instant winners
- Eliminate the useless or inappropriate ideas
- Sort similar ideas into groups and select best of each group
- Apply criteria to ideas that are instant winners and best of each group
- Submit the short-list ideas to reverse brainstorming (that is, in how many ways can this idea fail?)

KEY POINTS: THE ART OF BRAINSTRORMING

- The idea-generating process should be kept separate from the idea-evaluation process, because criticism tends to inhibit creativity.
- The brainstorming technique does work. But it requires good leadership from you to enable it to work.
- The secret is careful preparation. Analyse the problem so that you understand its dimensions and can present it coherently to the group. Make sure that the logistics of the meeting – flipchart and pens, cards or notepaper and a quiet room – are all in order.
- After the brainstorming, work ideas will have to be rough-sorted in terms of feasibility or practicality. Then you need to establish a process for further work to be done on them.
- Always remember to give some feedback on progress-towards-implementation to groups or individuals who have come up with new ideas.

> *Day-dreaming is thought's sabbath.*
> Amiel

CHECKLIST:
THE ART OF BRAINSTORMING

Answer Yes or No to the following:	Yes	No
Do you make use of the brainstorming technique when the situation or problem invites it?	☐	☐
If brainstorming doesn't seem to work do you always check to see if it isn't due to some failure in your own leadership?	☐	☐
Have you considered allowing another member of the team to lead an ideas-generating session?	☐	☐
Have you experimented with any variations on the basic brainstorming model?	☐	☐
Do group brainstorming sessions, in your experience, lead to much more individual creative thinking around the office or factory? If so, can you give examples?	☐	☐
Have you a shopping list of the main specific but open-ended problems that face you in your area of responsibility?	☐	☐
Does each first-line team leader have a similar shopping list?	☐	☐
Does your organization make sufficient use of project groups or creative thinking task forces?	☐	☐

13

TAKING GOOD IDEAS TO MARKET

Not so long ago cattle and sheep, even geese and turkeys, were herded to market in droves. Now they travel in trucks. Every farmer knows that it's no good hatching eggs and rearing plenty of chicks if they never reach the market. It is the same with new ideas. Innovation has to be market-driven – and driven to market.

Following that analogy for a moment, there should be a lively *internal market* for ideas in your organization. May I hasten to add that I'm *not* suggesting an extension of the internal market concept into intellectual property: one department selling ideas to another, a proliferation of internal invoices, and an accounting nightmare of lateral transactions. But your organization can and should be seen as analogous to a market in which ideas are, so to speak, up for sale.

In any market there are sellers and buyers. It's not a matter of the management being the buyers and the employees being the sellers. That perpetuates the old 'us' and 'them' division. In fact in a truly innovative organization everyone is potentially a seller and a buyer. As a manager you may want to 'sell' an idea to the workforce or the senior executive team or the trade union representatives. Equally, colleagues, trade union leaders or the chief executive may approach you as a 'customer' for their good idea.

> Sally Holdsworth works as a book editor for an international publishing company. At a dinner party a fellow guest mentioned that she was writing a children's book about moles, illustrated by her daughter. Sally looked at the fourteen chapters completed and felt that it showed exceptional talent.
>
> As she was not on the children's books side of the business she discussed it first with her publishing director. He regarded her judgement highly and promised to raise the idea at the next appropriate committee meeting. It passed that hurdle and then found support in the international division dealing with children's publications. Finance raised some objections on the artwork costs but Sally and her allies were insistent that the full original watercolours were integral to the project and must be used. When published, *Mole River* sold over one million copies and netted the company a handsome profit.

Notice that good ideas can come from so many different sources, some close at hand and some far removed. A comprehensive list of sources for new ideas in an industrial organization might include the following:

- Research and development
- Other specialist planning and research functions
- Directors and senior managers
- Quality circles
- Competitors
- Suppliers
- Customers
- External research establishments
- Government

As a general principle, people with a 'hands-on' involvement in any product or service – providing they have a modicum of interest in their work – will tend to have new ideas for doing it better. These will usually, but not always, be quite small or incremental improvements. But they are a vital part of the general process of innovation. Given encouragement and a listening leadership, this natural harvest of ideas can be increased dramatically. Any truly innovative organization should have 'buckets of ideas' available if it sets up some simple systems for lowering the buckets into the well and drawing them up.

Interest leads to ideas. In turn, the recognition of ideas by management leads to more job interests, greater involvement and deeper commitment. Even if – for good reasons that are explained – a team member's proposals are not acceptable, or if acceptable, cannot be implemented, there should be no loss of motivation. The important thing from the motivational perspective is the feeling of really being part of the enterprise, with a full share of responsibility in developing the quality of the product or service. Identification matters more than the fate of any particular suggestion.

Success in the outside markets of the world, then, depends upon the effective working of the *internal market* for ideas. The whole innovative process can be now shown in the simple model over the page, which illustrates the processes and key elements found in organizations that are successful innovators.

Much of this book has been concerned with generating new ideas – the larger, upper part of the light bulb. At the lower end of the model – the three rings on the base – are three stages or phases that ideas have to go through before they eventually contribute to the satisfaction or delight of a customer, one who values them enough to exchange good money for them:

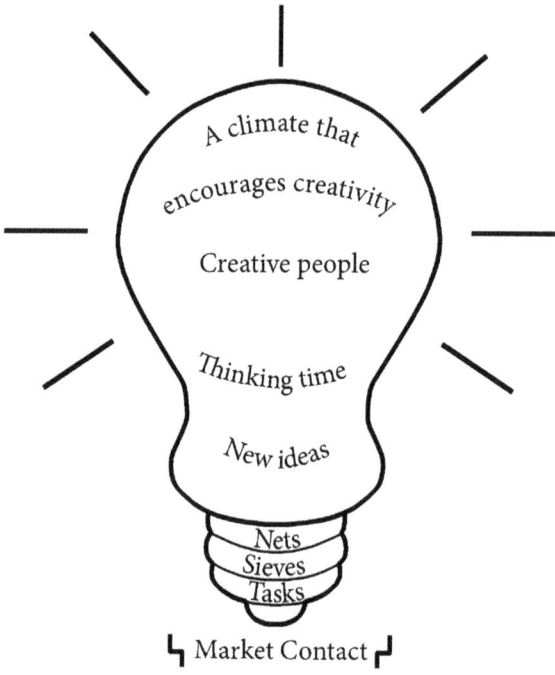

Successful Innovation

- **Nets**   The new ideas have to be netted, or harvested
- **Sieves** A sorting-out process is needed in order to separate the gold dust from the mud
- **Teams** Developing modifications or improvements, and bringing new products to market, always calls for a mix of talents, skills and knowledge

This model suggests that good new ideas are more likely to be forthcoming in an organization that encourages creative people and allows time for productive thinking. Needless to say, this requires leadership and vision to bring about and to sustain the right climate.

NETS: HARVESTING IDEAS

Ideas have to be harvested, like fish from the sea. What are the idea-catching nets or networks in your organization? The two principal ones are meetings and suggestion schemes.

Meetings

I know that meetings have acquired a bad name. As a manager working in an abysmally uncreative organization once said to me, 'Nothing is impossible here until it is sent to a committee.' But, given good leadership, meetings are your best way of netting ideas.

> **EXERCISE 14: Assessing the usefulness of meetings**
> Make a list of all the meetings you have attended in the last two weeks. How much time did each of them devote to netting new ideas as opposed to processing old ones?

The last chapter on brainstorming should have given you plenty of ideas on how to generate more ideas in meetings. Use fine mesh nets – leave the sorting out process until later on.

Suggestion schemes

The other way – apart from trawling nets – for catching fish in commercial quantities is by a variation of your individual rod-and-line method. With baited hooks at intervals along it, a line a mile or more in length is towed slowly behind the fishing vessel and then hauled in. The line method – with baited individual hooks – is a workable analogy for your company's suggestion scheme.

Ideas unlimited

In 1857 the Chance Brothers of Smethwick, surprised when their workers suggested ways of improving production and saving on materials, hit upon the idea of having a wooden box where such ideas could be posted. The scheme proved to be of immense worth to the firm and to the workers. It was the world's first suggestion scheme.

Suggestion schemes were not widely adopted, however, until this century. The Great Western Railway introduced one in 1913, and British Rail perpetuated the tradition. For example, a fitter who pointed out that there would be considerable saving if hydraulic pan jacks were re-sleeved with stainless steel liners received an award. Another employee suggested an improved design for shelves in the attendants' compartments in new sleeping cars. But for enthusiasm in offering suggestions, a railway employee with sixty years' service would be hard to beat. During his working life he offered no fewer than 30,800 ideas, thus earning himself a place in the *Guinness Book of Records*, if nothing else.

Since those early days, suggestion schemes have been renovated and renamed. Large organizations tend to have their own named schemes, such as Ideas Centres, Brainwaves, Bright Ideas, Winning Lines. This is all a part of giving them a fresh and positive image for marketing them internally. But the original concept remains unchanged: the invitation for new work-related ideas from individuals, with prizes or awards as an incentive for the successful suggestors. Suggestion schemes rest on a successful human formula:

- All ideas are welcome
- Everyone can put forward ideas

- We will listen to all ideas
- We realize that some ideas will not work
- But every idea deserves a cheer
- Ad a forum for ideas can be useful

This formula works successfully in many smaller companies without much organizing requirement, providing the managers give encouragement, opportunity, listening ears and some rewards. Like all incentive schemes, however, suggestion schemes in large organizations have to be properly administered. To run them fairly and effectively on this larger scale – with proper internal marketing, the necessary paperwork and the system of assessor committees – requires a commitment of management time and resources. This is the case even if you keep your scheme as simple as possible. Therefore there is clearly a cost/benefit judgement to be made if you want to introduce or update a suggestion scheme. If you do decide to introduce a new suggestion scheme or renovate an old one, make sure it's part of a coherent strategy.

A quick response to new ideas or suggestions is also essential. Knowledge of results is always motivating. Conversely, not knowing what has happened to your bright idea for months on end is extremely demotivating and demoralizing. The system must be such that participants know fairly soon if the organization is saying yes, no, or wait.

If the answer is no it is important to explain why in some detail. That requires either a personal letter or, preferably, a short meeting. Research suggests that people are not demotivated if their ideas are rejected, as long as the reasons for rejection are set out clearly and convincingly. Needless to add, we all stand in need of tact and diplomacy when our ideas are being rejected.

To enjoy success, success schemes need to be marketed

internally. Special events, publicity, newsletters and local newspaper or radio, together with a lively and compelling promotional booklet, are all ingredients in keeping the system alive and functioning well. Never expect schemes to go on working without maintenance, revision and re-inspiration.

SIEVES: SORTING IDEAS

Those assessment criteria in suggestion schemes introduce the next stage in the model – the sieves. You can separate wheat from chaff with a sieve. You can pan for gold with a sieve. Sifting is an applied form of that mental meta-function of valuing.

Ideas need to be evaluated rigorously at the right time. When given an idea or suggestion, Henry Ford used to ask three questions:

- Is it needed?
- Is it practical?
- Is it commercial?

These questions do have to be pressed home hard in commercial and industrial organizations. But, as we have seen, they should not be applied prematurely in the creative process. Sometimes ideas have to evolve quite far before any practical and commercial use becomes apparent. But tested they must be by others at various stages of their life history. The good ones are those that can jump the hurdles of criticism.

Testing or criticizing other people's new ideas – and being on the receiving end of that treatment – is often not a pleasant process. It can be downright demoralizing to the receiver. We have to learn to express our views with tact diplomacy.

Ridiculous!

'... A piece of paper just large enough to bear the stamp, covered at the back with a glutinous wash, which the sender might, by applying a little moisture, attach to the back of the letter...' An extract from the original proposal by Rowland Hill, inventor of the postage stamp. When the then postmaster general heard of it, he exploded: 'Of all the wild and visionary schemes I have ever heard of, or read of, this is the most extravagant!'

Remember that you are looking for a gold-bearing ore or gold dust, not finished gold cups or polished gold rings. So don't expect to find ideas that are instantly market-ready. Some new ideas will be dross. 'New ideas can be good or bad, just the same as with old ones,' remarked Franklin D. Roosevelt.

The process of sifting is much more like a busy trading floor, with buyers and sellers hard at work. You should be able to sell your ideas by presenting them effectively. But you are a buyer in that market as well. What matters is the intrinsic value of the idea. In your buyer role, put on your eyeglass and examine a new idea like a diamond, taking into account its colour, size and flaws. For it may, if adopted, cost your organization a lot of money.

TEAMS: SELLING YOUR IDEAS

'In the modern world of business,' said the advertising magnate David M. Ogilvy, 'it is useless to be a creative original thinker unless you can also sell what you create. Management cannot be expected to recognize a good idea unless it is presented to them by a good salesman.'

In other words, the onus is on you to persuade others that the proposed change is a good one, bearing in mind Henry Ford's three questions: Is it useful? Is it practical? Is it commercial? As money is the language of business, you have to be able to show that – at least in the middle term – the new idea or innovation will cut costs, add to profits or serve some other legitimate corporate interest. You sell ideas best by pointing out the benefits it will confer upon the 'buyer', whether he is an external customer or an internal member of the same organization as yourself.

Why it's not going to be easy

There is a natural opposition among men to anything they have not thought of themselves.

Barnes Wallis

If you do have a good idea, then be persistent in selling it. Colonel Sanders, who was made redundant at the age of about fifty, had only one key asset: a good recipe for a local dish. He made 1,100 calls to try to sell that idea, only to face 1,100 rejections. Wouldn't you have given up? But Sanders showed the necessary persistence that stems from belief in an idea. The very next company accepted his idea and Kentucky Fried Chicken was born.

But, given the realities of human nature, perseverance is as important for you as an intrapreneur – an innovator within an organization – as it is if you are an entrepreneur. Successful selling depends in large measure on creating interest. No farmer sows seeds into hard, frozen or unyielding ground. You have to prepare the ground for change.

Unless you can create some dissatisfaction with things as they are, you cannot induce a willingness to change.

EXERCISE 15: Selling your ideas

In the box below, write down an idea you'd like to have implemented in your organization and note down some of the supporting and opposing forces:

IDEA:	
FORCES AGAINST	**FORCES IN FAVOUR**
_____	_____
_____	_____
_____	_____
Now list some ways in which you can *reduce* the strength of these forces.	How can you *build* on these factors favouring the idea?

You cannot just come up with a good idea and expect someone to accept and develop it. You have to be prepared to push it all the way through to the market.

Major innovation should be planned in gradual stages, as part of a continuous process of adaptation to changing circumstances. It should not be a panic response to change that is now confronting an organization because yesterday that same organization failed to accept and respond to the requirements of change. Use the time available carefully to communicate about the need for change, experiment and review. 'Desire to have things done quickly prevents their being done thoroughly,' reflected Confucius. With innovation it is usually best to make haste slowly. Take change by the hand, before it takes you by the throat.

INDIVIDUAL CHARACTERISTICS OF INNOVATORS

- A clear vision of the objectives in mind – even if you are not clear initially how to go about achieving them.
- The ability to define the specific objectives and benefits of the emerging project.
- The skill to present the case for it persuasively and cogently.
- Support, not only from your superior(s) but also from colleagues and subordinates – you need to build an alliance in which everyone feels they are partners in a project that all are convinced is worthwhile.
- Courage – to take calculated risks and to face the breakers when difficulties or setbacks are encountered.
- The ability to motivate and inspire people into action, so that everyone contributes fully to the project and participates in the necessary decisions.
- The influence to mobilize support and resources in order to get things done.
- The ability to cope with interference or opposition to the project – such as criticism of details, tardiness, lack of corporate enthusiasm, requests or disputes over allocation of time and resources among projects.
- The willpower to maintain momentum, especially after the early enthusiasm for the project has declined and the team is involved in the hard work necessary.
- The determination to ensure that the whole team shares fully and fairly in the rewards of success.

KEY POINTS: TAKING GOOD IDEAS TO MARKET

- Innovation is more than having new ideas: it includes the process of successfully introducing them or making things happen in a new way. It turns ideas into useful, practicable and commercial products or services.
- Successful innovative organizations have nets for harvesting ideas, such as team meetings specifically for that purpose as well as supporting suggestion schemes. Ideas should come into those nets from many sources – competitors, customers and market research as well as from your staff.
- After preliminary sifting or evaluation, good suggestions still need to be sold in the internal market place of ideas. You should not only aim to be a good 'seller' of ideas but also an open-minded and potentially enthusiastic 'buyer' of ideas from your colleagues.
- Forming a project team with an effective leader is a critical step in the story of most significant innovations. An intrapreneur or 'product champion' often plays a key role as well, usually in association with a team.
- In overcoming the natural resistance to change, the inertia which tends to accumulate in any organization (regardless of the merit of particular ideas), it's useful to remember three proven strategies: recruit a senior sponsor, offer a pilot project or experiment and present innovation as incremental development.
- Without good leadership desired change will not happen in time. Leaders need both personal qualities, notably enthusiasm, and also professional skills to involve others in decision making and the management of change.

People support what they help to create.
Anon

SUMMARY

The end results of successful innovation include:

> **stimulated team members**
> **delighted customers**
> **profitable growth**

He who dares nothing need hope for nothing.
                              English Proverb

CHECKLIST:
TAKING GOOD IDEAS TO MARKET

Answer Yes or No to the following: Yes No

Do you operate an effective internal market for potentially innovative ideas in your organization? ☐ ☐

Do all teams now set aside time at some of their meetings in order to harvest and sift ideas? ☐ ☐

In the last twelve months have you been away with your team for twenty-four hours in order to constructively and creatively review progress and plan for the future? ☐ ☐

Do you involve customers and suppliers in the overall process of innovation? ☐ ☐

Identify one team or department that is outstandingly innovative. What are the three main reasons for its success?

    1. ………………………………………………………………

    2. ………………………………………………………………

    3. ………………………………………………………………

Is your organization's suggestion scheme working efficiently and fairly, in harmony with the rest of your strategy for harvesting ideas from your staff? ☐ ☐

Do they receive adequate recognition when the product or service comes to market, along with the whole project team? ☐ ☐

Are good ideas sometimes lost in your organization because they are not effectively presented? ☐ ☐

Can you think of any other route to long-lasting and internally generated profitable growth apart from innovation? ☐ ☐

APPENDIX:

Solutions to Problems

Six Matches puzzle, page 41

Again, ask yourself the reason why you couldn't solve it. Were you making an assumption or imposing a constraint that the puzzle must be solved in two dimensions? Breaking into three dimensions gives the elegant solution:

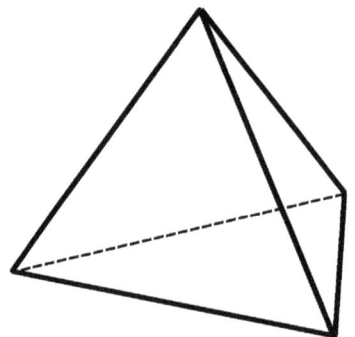

There is a second solution:

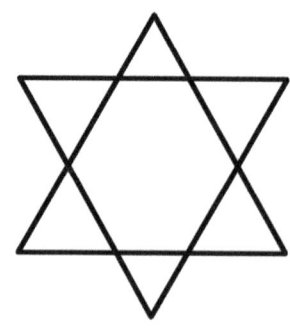

In so far as it involves putting matches on top of each other, it is a slight move away from two-dimensional thinking.

The puzzle was devised by a German psychologist called Karl Dunckner in 1924.

Quiz, page 50
1. Sculptor
2. Traveller in corks
3. Musician
4. Undertaker
5. Journalist
6. Veterinary surgeon
7. Television engineer

Answers to pages 55–58

1. Who Owns the Zebra?
The Norwegian drinks water.
The Japanese owns the zebra.
It is worked out as follows:

Front doors	Yellow	Blue	Red	Ivory	Green
Inhabitants	Norwegian	Ukrainian	Englishman	Spaniard	Japanese
Pets	Fox	Horse	Snails	Dog	Zebra
Drink	Water	Tea	Milk	Orange Juice	Coffee
Ice cream	Vanilla	Strawberry	Chocolate	Raspberry	Banana

2. The Swimming Pool
The dotted line represents the new pool twice the size:

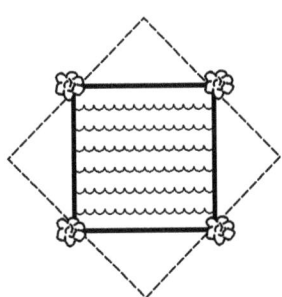

3. The Restaurant Meal
One of the women is the grandmother, and *her* two daughters are the mothers of the other *four* daughters. So there are seven of them altogether.

4. Relatives
The doctor in London is a *female* doctor – so she is the sister, not the brother, of the lawyer in Manchester.

5. Bottled Money
Push the cork *into* the bottle, and shake out the coin.

6. Farmer's Choice
The farmer arranges the twelve long hurdles in the shape of six joined-up triangles. Then he uses the six short hurdles to divide each triangle into two, like this:

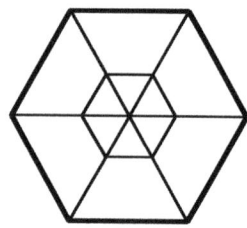

7. Prisoner's Escape
Harry piles the earth in a mound until it reaches the skylight.

8. Drinking Glasses
Pick up the middle one of the full glasses and, after pouring its water into the middle empty glass, put it back in its original position. This one helped you to assess your *mental flexibility*. The ability to *categorize your options* (push, pull, lift, slide, pour, etc.) is also involved here.

SOLUTIONS TO PROBLEMS

9. The Bicycles and the Fly

Each bicycle travels at 10 miles an hour, so the two will meet at the centre of the 20-mile distance in exactly one hour. The fly travels at 15 miles an hour, so at the end of the hour it will have gone 15 miles.

Many people try to solve this problem the hard way. They calculate the length of the fly's first path between bicycles, then the length of his path back, and so on for shorter and shorter paths. But this involves what is called the *summing of an infinite series*, and it is very complicated, advanced mathematics.

10. The Three Ties

Mr Brown has a black tie. Mr Black has a green tie. Mr Green has a brown tie.

Brown couldn't be wearing a brown tie for then it would correspond to his name. He couldn't be wearing a green tie because a tie of this colour is on the man who asked him a question. So Brown's tie must be black. This leaves the green and brown ties to be worn respectively by Mr Black and Mr Green.

Quiz, page 90

1. A young English designer named Carwardine approached the firm of Herbert Terry at the beginning of the 1930s with the proposal that they should build a desk light employing the constant-tension jointing principles found in the human arm. The company agreed, and the Anglepoise light was the result. From that time it has been in production, scarcely altered except for details and finishes.
2. Cats' eyes in the road.
3. Spitfires.

4. Clarence Birdseye took a vacation in Canada and saw some salmon that had been naturally frozen in ice and then thawed. When they were cooked he noticed how fresh they tasted. He borrowed the idea and the mighty frozen food industry was born.
5. They could have suggested the principle of independent suspension.
6. The burrowing movement of earthworms has suggested a new method of mining that is now in commercial production.
7. In the Royal Botanic Garden Edinburgh there is a plaque commemorating a flower that inspired the design of the Crystal Palace.
8. Sir Basil Spence, the architect of Coventry Cathedral, was flipping through the pages of a natural history magazine when he came across an enlargement of the eye of a fly, and that gave him the general lines for the vault.
9. Linear motors.
10. Ball-and-socket joints.
11. Magnifying glasses.
12. The arch. Possibly the Eskimos were the first to use the arch, in the construction of igloos.
13. Hollow steel cylinders.
14. Levers.
15. Bagpipes.
16. Wind instruments.

ACKNOWLEDGEMENTS

The author and publisher would like to thank the following for permission to reproduce material:

Gifford Pinchot, *Intrapreneuring* (Harper & Row, 1985)
Edward de Bono, *The Use of Lateral Thinking* (Jonathan Cape, 1967)
Roy Thomson, *After I Was Sixty* (Hamish Hamilton, 1975)
C. S. Forester, *Long Before Forty* (Michael Joseph, 1967)
Karl Popper, *Unended Quest* (Routledge, 1992)
Francis Crick, *What Mad Pursuit: A Personal View of Scientific Discovery* (Penguin, 1988)
G. Lakoff & M. Johnson, *Metaphor We Live By* (University of Chicago Press, 1986)
Gilbert Ryle, *On Thinking* (Blackwell, 1979)
Norman Thelwell, *A Millstone Round My Neck* (Methuen, 1983)
Donald T. Phillips, *Lincoln on Leadership* (Warner Books, 1992)
Charles Handy, *The Age of Unreason* (Century Hutchinson, 1989)
H. Rossotti, *Introducing Chemistry*, (Penguin, 1972)
James Michener, *Space* (Secker & Warburg, 1982)

Every effort has been made to trace all copyright holders but if any have been inadvertently overlooked the publishers will be pleased to make the necessary arrangement at the first opportunity.

INDEX

Page references in **Bold** denote complete chapters or sections
Works of literature are shown in *italics*.

Act of Creation, The (Koestler) 24
After I Was Sixty (Thomson) 27
Aha! experience 71
alternatives 11, 98–9
ambiguity **98–110**
  deciding between alternatives 98–101
  decision making funnel 98–9
  negative capability 101–2
  strategies for solving 103–5
Amiel, Henri F. 189
A Millstone Round My Neck (Thelwell) 118
analogy
  familiarization 92–3
  in language 87
  meaning of 88–9
  from nature 90, 91–3, 96
  in organizations 154, 191, 195–6
  as path to innovation 89–90
  practical use of 91–2
  stepping stones of **87–97**
analysis 22, 29, 52
  and depth mind principle 23–4, 26, 35

Anglepoise light 209
Art of Thought, The (Wallas) 69
Asimov, Isaac 120
assumptions
  assumptive thinking 43
  challenging 41–5, 84, 151, 181
  conscious 41–2
  failure to examine 15
  management 45
  present day 58
  unconscious 71, 84

Bacon, Francis 124, 174
Ballbarrow 49
Banner, John 91
barriers to creative thinking 14–15
Bates, H. E. 103
Benjamin, George 103–4
Bessemer, Sir Henry 49
Birdseye, Clarence 210
bisociation 24–5
Bonaparte, Napoleon 112
Boone, Daniel 102
Borden, John 147–8
Brady, William T. 159
Bragg, Sir Lawrence 82–3

brain
  see mind, working of
brainstorming 132, **177–90**
  chain-reaction in 185–6
  checklist 190
  definition of 178–9
  follow-up importance of 188
  Forced Connections 186
  ground rules for 181–4
  key points 189
  leading a session 184–8
  matrices use of 186–8
  in perspective 179–81
  problems suitable for 180–1
  silences in 185
Branson, Richard 172
Braun, Wernher von 169–70
Brenner, Sydney 167, 180
Brin, Sergey 8
British Rail 196
Brower, Charles 85
Brunel, Isambard Kingdom 69
Buchan, John 133
building on ideas 145–50
bureaucracy 112, 155

Cambridge 51, 82, 84, 149
Carroll, Lewis 123–4
Cavendish Laboratory, Cambridge 82, 149
champion, role of 7, 203
Chance Brothers 196
chance intrusions **61–7**
  chance inventions 61–4
  prepared mind 64
  recognizing 66
  serendipity 64–6
  welcoming 66–7
Chesterton, G. K. 122
Churchill, Sir Winston 146

Collier, John 133
colour relationships 33
Columbus, Christopher 65
communication
  and innovation 144, 150–1, 160, 201
  open lateral 210–17
  and teamwork 137, 167
  and unconcious mind 25, 32
Conran, Sir Terence 121–2
conscious and subconscious
  see mind, working of
Coventry Cathedral 76
creative thinking and creativity
  and ambiguity **98–110**
  and analogy **87–97**
  and assumptions 41–5
  and chance intrusions **61–7**
  and criticism 81–6
  and curiosity 112–14
  Decision Making Funnel 98–9
  and depth mind principle 23–31, 68–79
  explanation of 1–2, **3–18**, 24–45
  habits for success in **39–129**
  and ideas banking **111–29**
  and inspiration 105–8
  and listening 118–19
  and negative capability 101–2
  and observation 114–18
  paradox of 32–6
  and problems 52–60, 75–7, 167
  and reading 119–20
  and recording 123–5
  and relevance 45–52
  and sleep 75–7
  and suspending judgement **80–6**
  and teamwork 133–4, 136, 140–4, 149–50, 154–7
  and travelling 121–2

see also brainstorming; lateral thinking
Crick, Francis 51, 82–4, 149
criticism 81–6, 198
 and brainstorming 181–2, 184, 189
 constructive 149, 159, 166
Crossing the River 23
curiosity 111, 112–14, 118–19, 126, 141

Davy, Sir Humphrey 168
De Bono, Edward 9–13
 training in thinking 9–10
 writings 9
 see also lateral thinking
Decision Making Funnel 98–9
depth mind 23–31, **68–79**, 181, 185, 187
 and problem solving 35, 52, 75–7, 103–5
Desert Island Discs (Plomley) 76
Dickens, Charles 155
Disraeli, Benjamin 49
DNA 51
Dostoyevsky, Fyodor 174
dreams 7, 25, 75–7, 78, 165
 daydreams 20, 189
Drew, Richard G. 147–8
Drucker, Peter F. 160
Dyson, James 49–51

Ecclesiastes 5
Edison, Thomas 64–5, 106, 171, 182
Effective Decision Making (Adair) 21, 90, 98
effective thinking theory 21–3
Einstein, Albert 41, 42, 102, 109, 148

emotions 25–6, 115
entrepreneur 7, 29, 155, 161–2, 172
Epictetus (Roman philosopher) 172
executive stress 15, 17

fear of failure 15, 17
Flaubert, Gustave 116
Fleming, Sir Alexander 62
flexibility 160, 173, 208
following rules 15, 17
Ford, Henry 3, 11, 198, 200
Forester C. S. 30–1
France, Anatole 113
freezing oleum 19–20
Freud, Sigmund 25
Frost, Robert 72, 108
Fuller, Thomas 125

Gibbon, Edward 128
Gide, André 109
glass-making 74–5
Goethe, Johann von 44, 107, 121
Golding, William 76
Goldwyn, Sam 119
Goodyear, Charles 63
Google 8–9
Graham, Alex 32–4
Greene, Graham 106–7
groups, xii 13–14, 135–6, 142–3
 dominance of 22–3, 141
 and organizations 133, 151
 working in 82–3, 134, 145–6, 150
 see also brainstorming; teams and teamwork
guessing 42, 43, 44

habits for successful creative thinking
  Chance Intrusions, Welcoming **61–7**
  Depth Mind, Listening to **68–79**
  Ideas Banking **111–29**
  Nine Dots, Going Beyond **39–60**
  Stepping Stones of Analogy, Using **87–97**
  Suspending Judgement **80–6**
  Tolerating Ambiguity **98–110**
Handy, Charles 211
Harris, Jed 116
harvesting ideas 195–201
Hill, Rowland 199
Hobbes, Thomas 124
holism 29–31, 69, 80, 139, 153
Holmes, Oliver Wendell 147
Honda, Soichiro 12, 163–5, 174
Hunter, John 73

Ibsen, Hendrik 151
ideas
  and assumptions 41–5, 48–9
  banking **111–29**
  bisociation 24–5
  building on 139, 145–50, 155–7, 162
  and communication 150–1, 153–4
  generation of 47, 69, 72–5, 80, 108
  good, marketing of **191–205**
  harvesting 195–201
  preconceived 42, 43, 48–52
  see also brainstorming; creative thinking and creativity; lateral thinking; teams and teamwork

imagination 20, 25, 43–5, 66, 103
  see also creative thinking and creativity
innovative organisation, the **159–76**
inspiration 74, 105–8, 115, 121
  and the unconcious mind 24, 76
intelligence fields of 32–3
intrapreneur 7, 200, 203
intuition 26–7, 29, 70, 71, 125
  see also depth mind
inventions and discoveries 89–90, 104
  ancient 5
  ballbarrow 49
  Bessemer process 49
  bouncing bomb 52
  circumstances of 61–4, 104–5, 163
  cyclonic vacuum cleaner 50
  Google 8–9
  mirror galvanometer 63
  offset printing 62
  phonograph 182
  postage stamps 199
  saccharine 62
  Scotch Tape 147
irrelevance 46, 47, 65

Japan and the Japanese 123, 125, 163–5, 172, 173–4
jellyfish 30
Jenner, Edward 52–3, 73
Johnson, M. 88
Johnson, Samuel 112
judgement, suspending **80–6**

Kawamoto, Nobuhiko 165
Kawashima, Kiyoshi 165
Keats, John 72, 82, 101

Kennedy, John F. 168
Kentucky Fried Chicken 200
Kettering, Charles F. 156
Kipling, Rudyard 103, 125
Koestler, Arthur 24–5
Kume, Tadashi 165

Laboratory of Molecular Biology, Cambridge 51, 82, 149
Lakoff, G. 88
lateral thinking 13–16, 52
  De Bono's concept 9–10
  definition of 11–12
  see also creative thinking and creativity
leadership 29, 154–6, 194
  and change ix–x, 163–5, 203
  and communication 151, 166, 193, 195
  responsibilities of 82, 133, 134–5, 153
  and risk 168, 170–1, 173–4
  and teams 134–7, 143–4, 161–2, 166
  understanding 13–14, 46, 93–4
  see also brainstorming; management
le Carré, John 105
Lehr, Lewis W. 168
Leonardo da Vinci 5, 73, 75, 104, 113
light bulb model of innovation 193–4
Lincoln, Abraham 161–2
listening 29, 183, 193, 197
  and communication 151, 155
  to depth mind 35, **68–79**
  for ideas 118–19, 126, 150, 156–7
Livingstone, Dr David 138–9

Lloyd-George, David 97
Lloyd Webber, Andrew 172
logic and logical thinking 9–10, 11, 20
  and new ideas 47, 143
  over-reliance on 15, 17
Long Before Forty (Forester) 30–1
long-term perspective 172–3
Lowell, Amy 71–2
luck 61–4, 103

management 34–5, 43, 44–5, 85, 150–1
  attributes of 112, 153, 155, 160
  of innovation **133–58**, 161, 166, 168–71, 173
  and leadership 93–4, 135, 137, 166
  and marketing ideas 191–2, 197, 199–200
  see also brainstorming; leadership
marketing ideas 148, 153, 187, **191–205**
  strategy for 173, 199–201
Marks and Spencer 121
matrices, use of 186, 187–8
mechanical intelligence 33
mental road blocks 73–4
meta-functions of mind 21–5, 28, 35–6
  use of 77, 80, 181, 198
metaphors 23, 25, 87–8
Metaphor We Live By (Lakoff and Johnson) 88
Michener, James 170
Miller, Henry 35
mind, working of **19–36**
  see also brainstorming; creative thinking and creativity; depth mind; lateral thinking

mistakes **80–6**, 142, 154–5, 168–71
Morita, Akio 112
Murdoch, Iris 27–8, 84, 95
musical perception 33, 150

Napoleon Bonaparte 112
negative attitude 15
negative capability 101–2
newness 1, 6, 16
Newton, Isaac 42, 63
Nine Dots, Going Beyond **39–60**
numeracy 33

observation 63–4, 66, 73, 114–18, 126
offset printing 62
Ogilvy, David M. 199
Olivier, Laurence, Baron 116
On Thinking (Ryle) 113–14
OPB (Other People's Brains) 7
OPM (Other People's Money) 7
organizational climate 131–2, 166
Osborn, Alexander F. 178, 181, 182, 184

Paderewski, Ignacy Jan 106
Page, Larry 8
Paine, Thomas 148
paradox of creative thinking 32–5
Parker, Henry 4
Pasteur, Louis 43, 44, 64, 66
Pauling, Linus 177
perception 4, 6, 12, 46, 47
  and fields of intelligence 33
perseverance 74, 102, 171, 200
Phillips, Donald T. 162
Picasso, Pablo 115
Pilkington, Sir Alastair 61–2
Pinchot, Gifford 7

Plomley, Roy 76
Popper, Karl 44
prepared mind 64–7, 108
problem definition 6–18
  analysis 27, 35, **39–60, 61–7**
  see also brainstorming; depth mind
project groups 143–50, 167, **177–90**, 203
puzzles 39–41, 55–8

qualifications 51–2, 170
Quality Circles, xii 178, 192
questions 49, 54–5, 143, 175–6
  and brainstorming 185, 186, 187
  and ideas 153, 198, 200
quizzes 50, 90, 151

radar 91
reading, xv 111, 119–20, 127
received opinion 42
recording skills 111, 118, 123–6, 127–8
Relativity, General Theory of 41
relevance, span of 45–52
risks, taking
  in organizations 160, 167–70, 173–4
  willingness for 122, 154–5, 202
Rodin, Auguste 32, 104
Roosevelt, Franklin D. 199
Rossotti, Hazel 53
Royal Navy 112
Rubel, Ira W. 62
Rubens, Peter Paul 4
rules
  bending and breaking 44, 155, 181
  for brainstorming 181–4, 185, 186

following 15, 17, 179
  imposing 40, 44–5, 174
Ryle, Gilbert 113–14

Sanders, Colonel 200
Sartre, Jean-Paul 86
Schiller, Johann 80
Scotch Tape 147
Segovia, Andrés 109
self-criticism 18, 81
Seneca 94
serendipity 64–6, 123
Shakespeare, William 101, 105, 116, 118, 125
short-termism 172
Six Matches Puzzle 41, 206
sleeping on problems 75–9
Smuts, Jan 29
solutions to problems 206–10
'solvitur ambulando' 72
Souriau, Etienne 10
Space (Michener) 170
span of relevance 45–8, 58–9, 60
spatiality 33
specialization 49, 60
Spencer Penknives 4
Spence, Sir Basil 76, 210
Sperry, Roger 19
split brain theory 19–21
sponsor, role of 7, 203
Steele, Sir Richard 127
Steinbeck, John 97
Stevenson, Robert Louis 49, 77, 139
Stickton Lacey Ltd 137
suggestion schemes 195–7, 203, 205
suspending judgement 80–6, 148, 150, 153
  in brainstorming 179, 180–1, 182
Sutherland, Graham 114–15
synthesizing 21–5, 35–6
  and depth mind 68–9, 70–1, 77
  as quality of creativity 75, 106, 139, 141, 143

Tchaikovsky, Peter Ilyich 68
teams and teamwork ix–x, 150–1
  and ideas marketing 194, 199–201, 203
  leading 134–7, 145–50
  in management 155–6, 161–7, 171, 173
  in organizations xi–xii, 157, 160
  see also groups; leadership
technology 147
  in history 5, 73, 162
  innovation in 50–1, 66, 91
  see also inventions and discoveries
television set cases 91
Thelwell, Norman 118
The Three Princes of Serendip 64
The Use of Lateral Thinking (De Bono) 9
thinking 113–14, 119–25
  and ambiguity **98–110**
  analogically **87–97**
  analytically 52, 64–6
  and assumptions 41–5
  habits for successful **39–129**
  holistic 29–31, 69, 80, 139, 153
  laterally 9–16, 52
  and the mind **19–36**, 69–78, 141
  purposeful **3–36**
  and span of relevance 45–52, 58–9

thinking (*cont.*)
  see also brainstorming; creative thinking and creativity; lateral thinking
Thomas Aquinas St 3
Thomas Leslie 106
Thompson, William 63
Thomson of Fleet, Roy 1st Baron 65
Thomson, Roy 27, 31
Thoreau, Henry 28
3M 147–8, 168–9
Three Circles model 134–6
Training for Decisions (Adair) 21, 39
travelling 65, 121–2, 127
Trevelyan, G. M. 84
Trevor, William 113
Tull, Jethro 47–8
Turner, Tom 137

vaccination 52–3, 73
valuing 22–4, 27, 28, 35

  as quality of creativity 139, 141
  see also brainstorming; depth mind; judgement, suspending
verbal inntelligence 33
Victoria, Queen 106
vulcanization of rubber 63–4

Wallas, Graham 69, 70, 71
Wallis, Barnes 52, 141, 200
Walpole, Horace 64
Watson, James 51, 149
Watt, James 107
Weinstock, Arnold, Baron 91
What Mad Pursuit: A Personal View of Scientific Discovery (Crick) 51, 83, 149
Wheldon, Sir Huw 72
Wilde, Oscar 171
Wordsworth, William 24
Wright, Orville and Wilbur 171

Young, Edward 186

www.ingramcontent.com/pod-product-compliance
Ingram Content Group UK Ltd.
Pitfield, Milton Keynes, MK11 3LW, UK
UKHW041306180426
11947UKWH00009B/731